Coastal Research Library

Volume 34

The aim of this book series is to disseminate information to the coastal research community. The Series covers all aspects of coastal research including but not limited to relevant aspects of geological sciences, biology (incl. ecology and coastal marine ecosystems), geomorphology (physical geography), climate, littoral oceanography, coastal hydraulics, environmental (resource) management, engineering, and remote sensing. Policy, coastal law, and relevant issues such as conflict resolution and risk management would also be covered by the Series. The scope of the Series is broad and with a unique cross-disciplinary nature. The Series would tend to focus on topics that are of current interest and which carry some import as opposed to traditional titles that are esoteric and non-controversial.

Monographs as well as contributed volumes are welcomed.

Charles W. Finkl, Coastal Education & Research Foundation, USA Email: cfinkl@cerf-jcr.com

More information about this series at http://www.springer.com/series/8795

Risa Palm • Toby Bolsen

Climate Change and Sea Level Rise in South Florida

The View of Coastal Residents

Springer

Risa Palm
Urban Studies Institute, School of Public Health
Georgia State University
Atlanta, GA, USA

Toby Bolsen
Political Science
Georgia State University
Atlanta, GA, USA

ISSN 2211-0577 ISSN 2211-0585 (electronic)
Coastal Research Library
ISBN 978-3-030-32601-2 ISBN 978-3-030-32602-9 (eBook)
https://doi.org/10.1007/978-3-030-32602-9

This Springer imprint is published by the registered company Springer Nature Switzerland AG.
The registered company address is: Gewerbestrasse 11, 6330 Cham, Switzerland

Acknowledgments

The authors would like to thank Georgia State University for providing a semester research leave for Professor Bolsen, and for providing funding to support the acquisition of survey responses from Qualtrics, as well as funding to support graduate research assistance. We are grateful to both the Department of Political Science and the Office of the Provost for their support of this project. We thank Justin Kingsland for critical assistance with data collection, cleaning, analysis, and the creation of several figures that appear in the book. We appreciate the work of Xinyu Fu for creating the map of the study areas as well as numerous figures that appear throughout the book. We thank Justin Bolsen for creating a figure that we used in the book. We also thank Jeremy Craig for his careful review of the manuscript and suggested changes at several stages of its preparation. We thank the First Street Foundation for providing access to the FloodIQ.com site information. We are grateful for the helpful suggestions of several colleagues, including Donald Reitzes and David Greenland. Jeremy Kopkas not only provided editorial advice but also prepared the manuscript following the publisher's guidelines for electronic submission. Of course, any errors should be attributed to the authors.

Contents

List of Figures

List of Tables

Chapter 1
Introduction

Abstract In the United States, a large and influential minority of Americans does not accept that climate change has been exacerbated by human activity. But are residents of at-risk areas more aware of their vulnerability? In late 2018, about 1000 coastal residents of South Florida were surveyed to see if this population, one that is particularly at-risk of flooding associated with sea-level rise, high tides and high intensity hurricanes, was aware of risks to their own communities and their own homes. We also tested whether or not exposure to large-scale maps of likely flooding to the city or county in which the residents live would increase awareness of the reality of the threat of sea-level rise, and its potential impact on real estate values.

Keywords Sea-level rise · Climate change · South Florida

> Alice is a 45-year-old resident of Miami Beach. She graduated from the University of Miami and is employed in financial services. She and her husband bought their home in 2004, and with the exception of the 2008–2010 economic downturn, they have seen its value steadily rise. She voted for Donald Trump for president in the 2016 election and sees herself as moderately conservative. Although she has seen street flooding in Miami Beach during the time she has lived there, her own home has never been flooded. She has seen headlines such as "Is Miami Beach Doomed to Drown," but she has her doubts about whether or not the sea level is actually rising. She is not worried about sea-level rise and does not believe that her home is at risk.

> Edward is a 68-year-old resident of Clearwater, Florida. He has lived in Florida all his life and retired to his dream home on the coast and close to a golf course. He is a political independent and voted for Hillary Clinton in the 2016 presidential election. He attended college and is very interested in politics. But he doesn't believe all the hype about climate change or sea-level rise. It is his understanding that the sea level on the Gulf Coast of Florida will rise by one inch every 3 years at most. Since his home is at an elevation of 9 feet above sea-level, it would take 108 years before his home would be at sea-level. To him, the issue of sea-level rise is far from urgent.

Climate change has received an increasing amount of attention over the past decade, particularly with recent reports predicting unprecedented flooding, droughts, and wildfires for the United States, along with new record increases in global temperatures. With warmer ocean temperatures, coastal areas will experience more frequent

R. Palm, T. Bolsen, *Climate Change and Sea Level Rise in South Florida*, Coastal Research Library 34, https://doi.org/10.1007/978-3-030-32602-9_1

and intense storms as well as more flooding associated with sea-level rise. In South Florida, the effects of a combination of land subsidence and sea-level rise are already visible in the form of more frequent street flooding from high tides and from coastal storms. This low-lying area has the deadly combination of population concentrated on the coast, underlying geology that limits the drainage of flood waters, settlement built on land fill, and generally low elevations susceptible to storm surge.

Several scholarly studies have shown the negative impact of coastal location in Florida on housing values (Bernstein et al. 2018; Rao 2017). Newspaper articles declare that "Florida has more to lose with sea rise than anywhere else in the U.S." and that it is "about to be wiped off the map" (Harris 2018; Rush 2018). But are the residents of South Florida concerned? Are those who live at particular risk from sea-level rise aware of this risk? Do they seek information about their susceptibility to climate change? Do they support policies and laws to mitigate the pace and extent of climate change that would, in turn, slow sea-level rise? Answers to these questions are not only of general interest, but they are also key to our understanding of factors that will affect likely changes in individual behavior and public policy.

The issue of climate change has become so politicized in the United States in recent decades that scientific findings are often interpreted through the lens of political party or ideology. For example, several studies have shown that when those who oppose government regulation are presented with threatening findings about the increasingly likely impacts of climate change, their response is *not* more concern with the issue, but instead resistance and even doubt about the veracity of the science. Do findings such as these that have been noted for the country as a whole also hold for residents of areas that are clearly at risk?

Effective communication about climate change is particularly important in a representative democracy like the United States, because successful public policy can only be enacted and supported with the concurrence of the majority of the population and the advocacy of its leaders. Although surveys show that the majority of Americans accept the idea that human activity has accelerated climate change, a large minority does not, and this minority is not declining in numbers and influence. Further, some national leaders have stated publicly that they doubt the findings of climate scientists and therefore oppose legislation that would cut carbon emissions.

Sea-level rise is one of the effects of climate change that especially threatens the residents of South Florida. Scientists have documented a direct relationship between climate change and sea-level rise through the mechanisms of the expansion of water volume associated with ocean warming and the absorption of more solar energy that accompanies the melting of glaciers and ice sheets (NOAA 2019). Despite this, there is even less consensus about the existence and causes of sea-level rise.

One of the challenges for those seeking to generate greater consensus among the public for action is to find better ways to communicate scientific information about climate change. Numerous survey experiments have shown the impacts of exposure to different types of climate change messages varying, for example, in the ways they are presented or in who presents the information. However, little fieldwork has been done in settings where the local effects of climate change are particularly evident and are currently an immediate threat to property and lives. This is why we have focused *on this issue in this place:* we assessed responses to sea-level rise by residents of coastal Florida, an area within the United States that is likely to suffer the most severe damage over the relatively short-run. We also provide new information about the attitudes of this particular population concerning climate change mitigation policies including land use regulation and carbon taxes.

We surveyed almost 1000 residents of South Florida in late 2018 to see if exposure to large-scale maps of likely flooding would increase acceptance of the reality of human-exacerbated climate change, sea-level rise, and the linkage between the two. We also were interested in knowing whether those who received this kind of information were more likely to believe that their own communities were at risk, and that the values of their own homes might be reduced. Our empirical findings—for example, that provision of specific and relevant visual information caused respondents, particularly Republican respondents, to become *less* interested in the subject—are of urgent importance.

In the next two chapters, we will examine both the scientific foundations for climate change and sea-level rise, and also the politicization of this topic in the United States. We will then describe the South Florida study area, the general characteristics of our respondents, and present an analysis of an experiment in which residents were provided with detailed, visual information about the risk of flooding associated with sea-level rise. We focus on predictive factors that shape beliefs, policy support and behavior. We also provide some detail about five of the respondents, each of whom is convinced his or her home is not at risk despite clear evidence to the contrary. We conclude with an analysis of the types of policy measures that these respondents support that would reduce carbon dioxide emissions or short-term vulnerability.

References

Bernstein A, Gustafson M, Lewis R (2018, Forthcoming) Disaster on the horizon: the price effect of sea-level rise. J Financ Econ. SSRN: https://ssrn.com/abstract=3073842. https://doi.org/10.2139/ssrn.3073842

Harris A (2018) Florida has more to lose with sea rise than anywhere else in the U.S., new study says. In: Miami Herald. Available at: https://www.miamiherald.com/news/local/environment/article213092454.html. Accessed 18 July 2018

NOAA National Oceanic and Atmospheric Administration (2019) Is sea-level rising? National Ocean Service. Available at: https://oceanservice.noaa.gov/facts/sealevel.html. Accessed 4 June 2019

Rao K (2017) Climate change and housing: will a rising tide sink all homes? Zillow. Available at: https://www.zillow.com/research/climate-change-underwater-homes-12890/. Accessed 16 July 2018

Rush E (2018) Rising seas: Florida is about to be wiped off the map. In: The Guardian. Available at: https://www.theguardian.com/environment/2018/jun/26/rising-seas-florida-climate-change-elizabeth-rush. Accessed 26 June 2018

Chapter 2
The Science of Climate Change and Sea-Level Rise

Abstract Although there is natural variability in climate when measured over millions or billions of years, the speed of climate change has accelerated as a result of human activity. Globally, the impacts of climate change include an increase in the annual mean surface air temperature and changes in rainfall patterns, with some regions getting more rainstorms and others becoming much drier. At a global scale, these rapid changes could result in regional food shortages triggering a massive amount of human migration, putting stress on food production as well as increasing political instability. In Florida, the impacts of climate change are likely to include flooding associated with sea-level rise, an increase in invasive species, damage to the coral reefs, and increased numbers of damaging hurricanes.

Keywords Atmospheric models · Carbon dioxide · Sea-level rise · South Florida · Fourth National Climate Assessment

Is the climate actually changing? And if it is, does this affect the sea-level? And how will these changes affect the South Florida region? To understand the context for our study, we will review how the climate system works, how we know that the climate is changing, natural and human forces changing the climate, how climate change affects sea-level rise, and the likely impacts of climate change and sea-level rise on South Florida.

2.1 How Does the Climate Work?

Elements of the climate system include the atmosphere, the surface of the land, the surface of the water and ice cover, as well as vegetation and human-made changes to the landscape, such as pavement and buildings. The interaction of these elements to form climate at a particular place is caused by the spinning of the earth on its axis, and the rotation of the earth around the sun, as well as the amount of solar energy released by the sun.

R. Palm, T. Bolsen, *Climate Change and Sea Level Rise in South Florida*, Coastal Research Library 34, https://doi.org/10.1007/978-3-030-32602-9_2

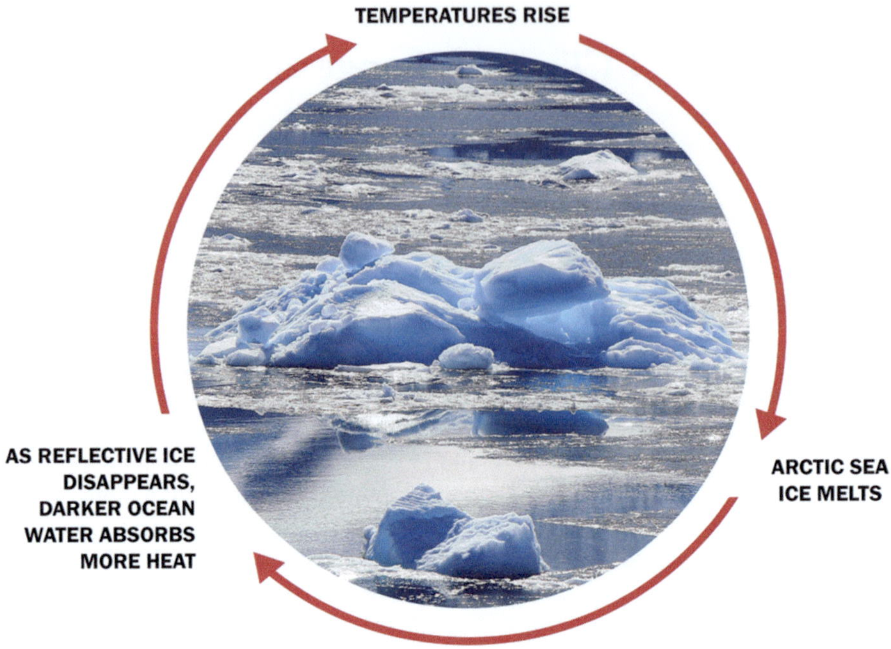

Fig. 2.1 Feedback loop: increased surface temperatures, sea ice melts, more solar energy absorbed

The earth's atmosphere traps heat from the sun. When the amount of radiation coming to the earth's surface is balanced with the amount that is lost into space, there is no net change in the earth's temperature. However, when more particles, water vapor, and gases such as carbon dioxide are in the atmosphere, more energy is trapped, and temperatures rise in the lower layers of the atmosphere. Further, because the climate is a complex system, changes in one part of the system can have large impacts, both direct and indirect on other parts of the system (Somerville et al. 2007; National Research Council 2010; Dessler 2016; Aguado and Burt 2014). For example, when water vapor, carbon dioxide, or other greenhouse gases trap the re-emission of solar energy previously absorbed at the earth's surface, they concentrate more warmth near the surface of the earth. This process can cause the light-colored ice and snow to melt, creating new water surfaces that are darker. This darker surface cover in turn causes more of the sun's heat to be absorbed rather than reflected, causing a self-perpetuating feedback loop of increased temperatures, more ice melt, more water, and further increases in temperatures (Fig. 2.1).

2.2 Evidence That the Climate Is Changing

In order to know whether or not the climate is changing, it was important to develop accurate instrumentation to measure temperature and rainfall, and to collect this information over a long period of time. Evidence of ancient climates is found in tree rings, layers of sedimentary rocks, ice cores, and ocean sediments. Evidence of more contemporary climate change is collected even more precisely from what are now automatic readings at thousands of stations both on land and on the ocean, as well as from earth-orbiting satellites (Peterson and Vose 1997). These very accurate data are available for the past 60 years, with some databases containing even longer histories. Evidence from this detailed and accurate climate monitoring shows overwhelming evidence of recent and rapid global warming-roughly ten times faster than the average rate of ice-age-recovery warming (National Research Council 2010).

In the past, the Earth has been both much colder and also much warmer than it is today. Both natural forces and human activity have changed the climate. Natural forces affecting the climate include variations in the amount of energy radiated by the sun, changes in the orbit of the earth around the sun, change in the angle at which the earth tilts on its axis, changes in the composition of the atmosphere (such as from volcanic eruptions), and the movement of continents through plate tectonics. Over the past 3 million years, Earth's climate has had a series of glacial and interglacial periods related to changes in the earth's orbit, also called Milankovitch cycles. In addition, over very long time-frames, the amount of solar radiation has varied with the luminosity of the sun as well as with the number of sunspots. Climate has therefore shown natural variability, particularly when understood over millions or billions of years. What has altered is the *speed* of climate change, and the effects of *human activity* on the climate. Alternative explanations for this pace of change have been sought and analyzed, and climate scientists have concluded there is one dominant cause: human activity related to the emission of carbon dioxide in the atmosphere.

How has human activity caused climate change? Evidence from ice cores in Greenland and Antarctica show that the amount of carbon dioxide and other greenhouse gases in the atmosphere has risen since the start of the Industrial Revolution in 1750. Greenhouse gases are defined as those that absorb infrared radiation and re-emit this energy into the atmosphere. The primary greenhouse gas is carbon dioxide, but the term also includes water vapor, methane, nitrous oxide and ozone (IPCC 2015).

The release of carbon dioxide into the atmosphere has been caused primarily by the burning of fossil fuels. Since we know that there has *not* been an increase in solar output over the past 30 years, and since the physical mechanism of the impact of increases in carbon dioxide on surface temperatures is well understood, the conclusion is one on which the overwhelming majority of scientists agree: that most of the warming of the earth observed in the past 50–70 years must be attributed to human activity.

2.3 Projected Impacts of Climate Change

At a global scale, the effects of climate change will differ significantly by region. For example, warming is projected to be greater over the continents than over the oceans and will be greater at the polar regions of the world. Figure 2.2 portrays the

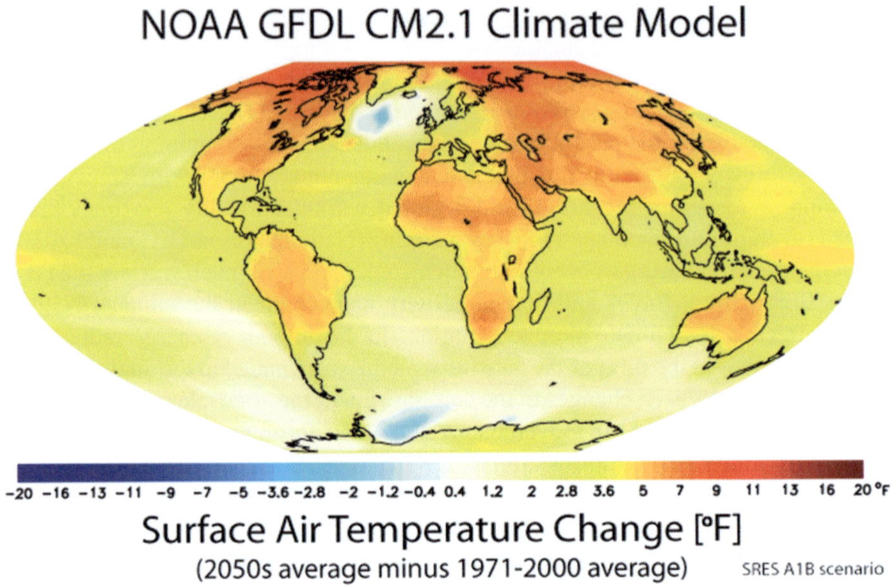

Fig. 2.2 Projected increases in surface air temperature to the mid-twenty-first century

projected change in annual mean surface air temperature from the late twentieth century (1971–2000 average) to the middle twenty-first century (2051–2060 average). The change is in response to increasing greenhouse gases and aerosols based on a "middle of the road" estimate of future emissions (GFDL 2019).

Rainfall will also change in amount, intensity, and location, with some areas getting more downpours and others becoming much drier (Fig. 2.3). This will result in more severe floods in some areas and droughts in others, affecting the ecosystem, agriculture, and human settlement.

These changes in both temperature and precipitation will have further impacts on both the climate system itself and on human life. First, a decrease in snow cover will affect the climate system because smaller areas of the earth will directly reflect solar energy back into the atmosphere, resulting in a greater accumulation of solar energy at the earth's surface. Second, decreased snowpack will also mean a reduction in the amount of water in rivers and streams on which major cities are dependent. This reduction of water in the rivers will also have major impacts on, for example, riverine ecosystems, river navigation, and the availability of hydroelectric power.

Oceans will become increasingly acidic as carbon dioxide is taken up by marine organisms and combined with water to become carbonic acid. This acidification can damage coral reefs, and change the ecosystem for a variety of fish, shellfish, and other resources on which humans depend.

As growing seasons change in length, soils will become drier in some areas, and the human agricultural system will be forced to adapt or even relocate. An increase in heatwaves, drought, and severe storms could also unleash pests and disease in

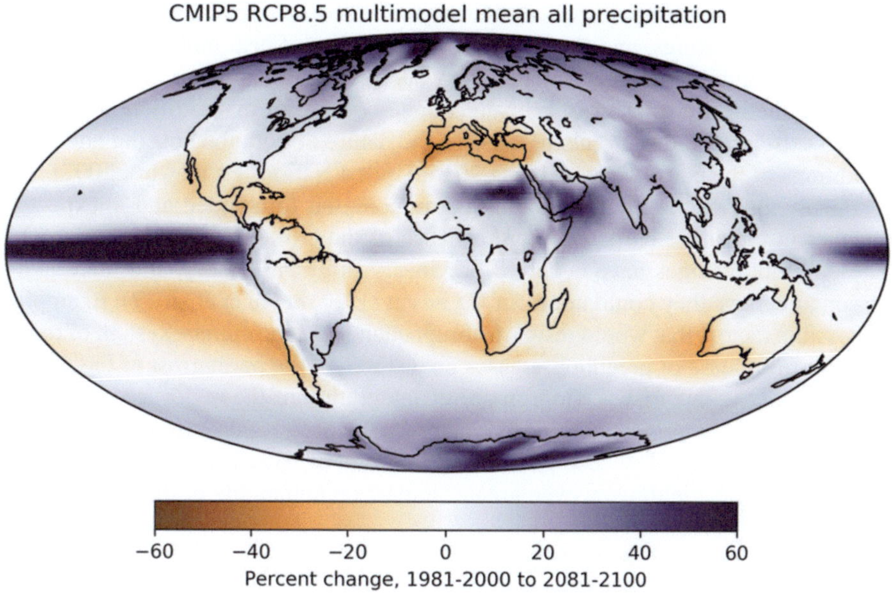

Fig. 2.3 Projected changes in rainfall 1981–2000 to 2081–2100. CMIP5 RCP8.5 multi-model average percent change in total precipitation (rain and snow) between 1981–2000 and 2081–2100. Uses one run for each model, 38 models total. (Data from KNMI Climate Explorer; map by Carbon Brief. https://www.carbonbrief.org/explainer-what-climate-models-tell-us-about-future-rainfall. https://web.archive.org/web/20160627173254/https://www.gfdl.noaa.gov/visualizations-climate-prediction/)

areas unaccustomed to such threats to agricultural production. Similarly, forests will be affected by changes in temperature and precipitation, as well as the introduction of new pests and diseases in regions where they previously had not thrived.

Climate change will also negatively affect public health. With increased extreme temperatures, more people, particularly those for whom air conditioning is not available, will suffer from heat-related stress. With increases in extreme storms such as hurricanes and floods, more people will be at risk. Added to all of this is the risk of infectious diseases adapting to areas where they had previously not been endemic, including mosquito-borne malaria and dengue, and tick-borne Lyme disease. Warmer temperatures will also mean that food will spoil more easily and pathogens in water will be able to reproduce more quickly, making the drinking water unsafe.

Societal impacts of climate change are also of concern. Human migration will increase both from drought-stricken areas that no longer support the production of food, and from areas that are being inundated from sea-level rise, bringing stress to neighboring political entities, or even sparking wars. The melting of Arctic ice and the increasing navigability of this region raises new potential conflicts about control of the waters for mining and fishing.

Poorer countries are likely to suffer more from climate change than richer countries because of (1) the dependence of a greater proportion of the population on agriculture, fishing, or forestry, all of which are susceptible to adverse effects of warming temperatures, and (2) the lack of availability of heat mitigation measures such as air-conditioning or access to fresh water. In addition, poverty creates vulnerability to the health impacts of climate change because of malnutrition, and exposure to extreme weather events because of the nature of both the housing and work conditions of the poor (Levy and Patz 2015; United Nations Framework 2007).

Finally, one of the greatest threats is the potential for sudden and unexpected changes in the entire climate system because of the crossing of a "tipping point." These abrupt changes have occurred previously in global history, and there is no reason to believe they could not recur (Lyons et al. 2018; Xu et al. 2017). Examples of possible factors that would result in abrupt changes in the climate system of today include: (a) release of greenhouse gases now stored in the permafrost of the Arctic, accelerating the rate of warming; or (b) disintegration of the major ice sheets and glaciers, releasing large quantities of water into the ocean and accelerating the process of sea-level rise. It is also possible that several of these processes could occur at the same time, yielding an even greater impact on human life or the environmental system.

2.4 Climate Change and Sea-Level Rise

Using a combination of satellite altimeter data with the more traditional measurement of tide gauges, scientists have concluded that the sea level is rising globally, and that the *rate* of sea-level rise is likely to increase (Nerem et al. 2018; Hay et al. 2015). The two major contributors to sea-level rise are ocean warming, with the concomitant expansion of the ocean volume, and ice loss to sea (Meier et al. 2007). The causal links between carbon emissions, the melting of Arctic ice, and sea-level rise show a time-lag and are also long-lasting (Clark et al. 2018; Moon et al. 2018). It is therefore clear that anthropogenic climate change will have a serious impact on sea levels and on human habitation of coastal areas for a long time to come. The connection between the phenomena of climate change and sea-level rise is so clear that a recent NASA publication on "FACTS" states that "sea-level rise is caused primarily by two factors related to global warming: the added water from melting ice sheets and glaciers and the expansion of seawater as it warms" (NASA 2019).

2.5 Implications of Climate Change for Sea-Level Rise in South Florida

Sea-level rise has become apparent in Florida. A 2010 report of the Florida Oceans and Coast Council noted that "sea-level rise is not a science fiction scenario but a reality … and the time is coming when the state will be simultaneously and continuously challenged by all of these effects" (Florida Oceans and Coast Council 2010). It concluded that "the long-term extent and severity of oceanic or coastal effects caused by climate change including sea-level rise ultimately depend on how rapidly humanity can eliminate human sources of carbon dioxide and other greenhouse gases entering the atmosphere at harmful levels, now and in the future" (Florida Oceans and Coast Council 2010).

In the same vein, immediately upon the release of the Fourth National Climate Assessment in November 2018, much of the local press outlined the impacts of climate change for Florida. One such article described five such impacts: the spread of illnesses such as the Zika virus, an increase in invasive species, damage to the coral reefs, increased numbers of damaging hurricanes, and increased rainfall (Shaffer 2018). Many other news reports focused on the relationship between climate change and sea-level rise.

Just as temperatures have fluctuated over the centuries, so has the level of the sea. Over the past 20,000 years, the sea level has risen about 121.92 meters (400 feet) (National Research Council 2010). Estimates show that the present global sea level was reached about 6000 years ago and has remained relatively constant until recent decades. Changes in sea level are caused by three factors. The first is the warming and cooling of the ocean, with associated expansion or contraction in its volume. When the ocean warms, expansion is greatest in the warm waters of the tropics and also near the surface. The second factor is the melting of ice that has been stored on land in the form of glaciers and ice sheets. A third related set of factors that can contribute to sea-level rise include the loss of wetlands through drainage and settlement, loss of forested lands that would otherwise capture carbon dioxide, and the mining of groundwater for domestic, agricultural, or industrial use.

In South Florida, the advance of ocean waters onto the land surface is caused by a combination of rising seas and sinking land surface. The rise in the sea level is part of general world sea-level rise, but also factors specific to South Florida such as ocean circulation patterns (Valle-Levinson et al. 2017). The land surface has also sunk as a result of post-glacial rebound in other parts of the continent, and because of the extraction of groundwater (Karegar et al. 2017; Piecuch et al. 2018; Williams 2016). Post-glacial rebound refers to the rise in the land surface that follows the melting of glacial ice. Glaciers depress the land masses, and when they shrink or disappear, the land rises or rebounds to previous elevations. The impact on Florida and other parts of the eastern part of the United States is a reduction in elevation as other parts of the continent that were formerly covered by glaciers rise with the removal of the ice sheet.

With an increase in ocean temperature and the melting of some of the earth's ice sheets and glaciers, both expanding the volume of ocean waters, South Florida faces an existential threat. In the coming chapters, we will explore how South Florida residents are coping with this threat.

References

Aguado E, Burt J (2014) Understanding weather and climate, 7th edn. Pearson, New York
Clark P, Mix A, Eby M, Levermann A, Rogelj J, Nauels A, Wrathall D (2018) Sea-level commitment as a gauge for climate policy. Nat Clim Chang 8:653–655. https://doi.org/10.1038/s41558-018-0226-6
Dessler A (2016) Introduction to modern climate change, 2nd edn. Cambridge University Press, Cambridge
Florida Oceans and Coastal Council (2010) Climate change and sea-level rise in Florida. Tallahassee. Available at: https://floridadep.gov/sites/default/files/Climate%20Change%20and%20Sea-Level%20Rise%20in%20Florida_1.pdf
Geophysical Fluid Dynamics Laboratory (2019) Data visualizations – climate predictions. Available at: https://www.gfdl.noaa.gov/visualizations-climate-prediction/. Accessed 30 Apr 2019
Hay C, Morrow E, Kopp R, Mitrovica J (2015) Probabilistic reanalysis of twentieth-century sea-level rise. Nature 517(7535):481–484. https://doi.org/10.1038/nature14093
Intergovernmental Panel on Climate Change (2015) IPCC Data Distribution Center. Available at: https://www.ipcc-data.org/guidelines/pages/glossary/glossary_fg.html
Karegar M, Dixon T, Malservisi R, Kusche J, Englehart S (2017) Nuisance flooding and relative sea-level rise: the importance of present-day land motion. Sci Rep 7(11197). https://doi.org/10.1038/s41598-017-11544-y
Levy B, Patz J (2015) Climate change, human rights, and social justice. Ann Glob Health 81(3):310–322. https://doi.org/10.1016/j.aogh.2015.08.008
Lyons S et al (2018) Palaeocene–Eocene thermal maximum prolonged by fossil carbon oxidation. Nat Geosci 12:54–60. https://doi.org/10.1038/s41561-018-0277-3
Meier M, Dyurgerov M, Rick U, O'Neel S, Pfeffer WT, Anderson R, Anderson S, Glazovsky A (2007) Glaciers dominate eustatic sea-level rise in the 21st century. Science 317(5841):1064–1067. https://science.sciencemag.org/content/317/5841/1064. https://doi.org/10.1126/science.1143906
Moon T, Ahlstrom A, Goelzer H, Lipscomb W, Nowicki S (2018) Rising oceans guaranteed: Arctic land ice loss and sea-level rise. Curr Clim Chang Rep 4(3):211–222. https://doi.org/10.1007/s40641-018-0107-0
NASA (2019) FACTS: Sea-level. Available at: https://climate.nasa.gov/vital-signs/sea-level/. Accessed 13 Apr 2019
National Research Council (2010) Advancing the science of climate change. The National Academies Press, Washington, DC. https://doi.org/10.17226/12782
Nerem R, Beckley B, Fasullo J, Hamlington B, Masters D, Mitchum G (2018) Climate-change-driven accelerated sea-level rise detected in the altimeter era. Proc Nat Acad Sci U S A (Cazenave C (ed)) 115(9):2022–2025. https://doi.org/10.1073/pnas.1717312115
Peterson T, Vose R (1997) An overview of the global historical climatology network temperature database. Bull Am Meteorol Soc 78(12):2837–2849. https://doi.org/10.1175/1520-0477(1997)078<2837:AOOTGH>2.0.CO;2. ISSN 1520-0477
Piecuch C, Huybers P, Hay C, Kemp A, Little C, Mitrovica J, Ponte R, Tingley M (2018) Origin of spatial variation in US East Coast sea-level trends during 1900–2017. Nature 564:400–404. https://doi.org/10.1038/s41586-018-0787-6
Shaffer C (2018) Trump administration climate report: 4 ways Florida could be impacted by climate change. In: Treasure Coast Newspapers. Available at: https://www.tcpalm.com/story/news/2018/11/23/trump-administration-climate-report-4-ways-florida-climate-change-impact/2094393002/. Accessed 25 Nov 2018

Somerville R, LeTreut H, Cubasch U, Ding Y, Mauritzen C, Mokssit A, Peterson T, Prather M (2007) Historical overview of climate change. In: Solomon S, Qin D, Manning M, Chen Z, Marquis M, Averyt KB, Tignor M, Miller HL (eds) Climate change 2007: the physical science basis. Contribution of working group I to the fourth assessment report of the intergovernmental panel on climate change. Cambridge University Press, Cambridge

United Nations Framework Convention on Climate Change (2007) Climate change: impacts, vulnerabilities and adaptation in developing countries. UNFCCC, New York. Available at: http://unfccc.int/resource/docs/publications/impacts.pdf

Valle-Levinson A, Dutton A, Martin J (2017) Spatial and temporal variability of sea-level rise hot spots over the eastern United States. Geophys Res Lett 44(15):7876–7882. https://doi.org/10.1002/2017GL073926

Williams C (2016) History of flushing Lake Okeechobee dates back to 1800s. In: News-Press. Available at: https://www.news-press.com/story/life/2016/07/17/history-flushing-lake-okeechobee-everglades-cynthia-williams-tropicalia/86930340/. Accessed 2 Dec 2017

Xu Z, Knorr G, Lohmann G, Barker S (2017) Abrupt North Atlantic circulation changes in response to gradual CO2 forcing in a glacial climate state. Nat Geosci 10:518–523. https://doi.org/10.1038/ngeo2974

Chapter 3
A Partisan and Polarized Issue in the United States

Abstract Climate change has become a politically polarized issue within the past 30 years, as interest groups and certain political leaders sought to dispute the growing scientific consensus about its causes and impacts. This chapter synthesizes a large body of survey, experimental and methodological literature that places the empirical study of South Florida in context. Past survey research has shown conclusively that party identification and ideology are the strongest predictors of climate change beliefs of Americans. Other predictors that are less consistent include demographic characteristics, cultural worldviews and personal experience. Survey and laboratory research has been directed at understanding the processes involved in accepting or denying messages about climate change. Among the findings are that strategically framed messages can shift opinion, that a belief in scientific consensus about climate change may increase acceptance of its reality, that prior beliefs, group identities and cultural worldviews moderate the acceptance of climate change information through motivated reasoning, and that best practices involve describing climate change as a personal risk, using social group norms to convince skeptics, and emphasizing social consensus on the issue. Prior research suggests that a message about environmental risk that is local and specific will be relatively more effective, particularly when the immediate threat is already visible.

Keywords Motivated reasoning · Framing · Partisanship · Public opinion · Consensus · Polarization · Risk communication

Climate change has become a polarized issue in the United States. The purpose of this chapter is threefold: first, to demonstrate the degree of polarization around the idea of climate change; second, to review some of the demographic, social, and personal factors that predict climate change beliefs and how they operate; and third, to review some of the studies that have addressed strategic messaging aimed at changing climate beliefs. All of these themes combine to describe the response of residents of the South Florida coast to both climate change and sea-level rise.

R. Palm, T. Bolsen, *Climate Change and Sea Level Rise in South Florida*, Coastal Research Library 34, https://doi.org/10.1007/978-3-030-32602-9_3

3.1 Polarization of Climate Change in the United States

> As Hurricane Irma barreled toward Florida across the Atlantic Ocean in September 2017, right-wing radio personality Rush Limbaugh suggested that U.S. media coverage of hurricanes approaching his home state was part of a liberal conspiracy intended to convince the public that climate change is real. Limbaugh, a resident of Palm Beach, stated on his program on September 5 that: "All you need is to create the fear and panic accompanied by talk that climate change is causing hurricanes to become more frequent and bigger and more dangerous, and you create the panic, and it's mission accomplished, agenda advanced" (Holan 2017). Hurricane Irma ended up being the most powerful hurricane in the Atlantic in recorded history with a death toll of 129 and total damage of $50 billion (Amadeo 2019).

> In November 2018, President Donald Trump dismissed the National Climate Assessment (NCA) report's conclusion that the U.S. would suffer up to $500 billion per year in economic losses by the end of the twenty-first century as a result of climate change (Irfan 2018). President Trump publicly stated that many scientists have "a very big political agenda" and that "As to whether it's manmade or not and the effects you're talking about are there, I don't see it –not nearly like it is…" The NCA report was produced by 300 expert scientists across 13 federal agencies including NASA and the U.S. Defense Department, and it is required by a U.S. law that has been in place since 1990.

As these examples make clear, influential conservative Republican leaders doubt that climate change is an immediate danger (also see, Nazarvan 2017; Palmer 2018; Shear 2017). In the United States, this division of opinion is reflected in different views within the general public as well.

There are two aspects to this disagreement about climate change. The first is whether or not the climate is changing, and the second has to do with the causes of such change, if it exists. Despite an overwhelming scientific consensus that human action is the dominant cause of accelerated climate change (Cook et al. 2016), a significant percentage of Americans do not believe that climate change is occurring (30%) or that it is human-caused (42%) (Leiserowitz et al. 2018; also see, Egan and Mullin 2017; Weber and Stern 2011). Partisanship divides opinions on both of these issues. Public discourse on climate change continues to be characterized by partisan and ideological divisions in the U.S. (Dunlap et al. 2016; Feldman et al. 2012; Hamilton 2016; Hart et al. 2015a, b). Democrats and liberals tend to believe climate change is occurring and that human activity is the dominant cause, but many Republicans and conservatives remain skeptical.

Some of the survey results have confounded the issue because of the interchangeable use of "climate change" and "global warming" in the survey questions themselves (Villar and Krosnick 2011; Schuldt et al. 2011; Schuldt et al. 2015). Schuldt et al. (2011) showed that belief that climate change is happening is higher among Republicans when the phenomenon is labeled "climate change" as opposed to "global warming." Schuldt et al. (2015) experimentally manipulated whether the issue was described as "climate change" or "global warming" to assess whether this semantic difference affected individuals' perceptions regarding (1) if it actually exists, (2) whether scientists agree about its existence, and (3) whether there is support for an emissions reduction climate-mitigation policy. Democrats were unaffected by the label "climate change" or "global warming"; however, Republicans were more likely to acknowledge that climate change is happening, that scientific consensus exists, and to support mitigation polices when the issue is was labeled "climate change" as opposed to "global warming" (also see, Schuldt et al. 2017).

A nationally representative survey fielded by the Pew Research Center in late March and early April of 2018 of 2541 adults in the U.S. revealed widespread disagreement concerning the existence of *global warming* and its main causes (Pew Research Center 2018). Respondents were asked to choose whether Earth is getting warmer primarily "because of human activity," "because of a natural pattern," or that "there is no solid evidence of any warming." Only 18% of respondents in the sample who identified as "Conservative Republicans" stated that that Earth is warming primarily as a result of human activity, and 36% of this group selected the option that there is "no solid evidence of global warming." In contrast, 83% of "Liberal Democrats" reported that Earth is getting warmer primarily because of human activity, and only 2% of this group saw no solid evidence of its existence.

Surveys have shown that Americans are also polarized over support for specific public policies that seek to address the problem. The 2018 Pew survey found that 94% of liberal Democrats, versus 57% of conservative Republicans, believed that "restrictions on power plants … would make a difference in reducing the effects of *global climate change*" (Pew Research Center 2018). Similarly, 89% of liberal Democrats said the same about the effects of "tougher fuel efficiency standards," compared to only 46% of conservative Republicans. Widespread partisan gaps were also reported with respect to support for polices such as providing a "tax incentives to encourage businesses to reduce carbon," "tax incentives for people to drive hybrid and electric cars," and support for the expansion of fossil fuel energy developments in the U.S (Pew Research Center 2018).

But this extreme partisan polarization is relatively recent. In 1988, the United Nations created the Intergovernmental Panel on Climate Change (IPCC) to study the impacts of climate change. NASA scientist James Hansen testified before Congress that global temperatures were increasing as a result of the burning of fossil fuels for energy. That same year, the country experienced an intense heat wave and drought, generating a substantial amount of media coverage about the "greenhouse effect" (Bolsen and Shapiro 2018; Krosnick et al. 2000; Ungar 1992).

In 1988, Republican George H.W. Bush campaigned for the presidency promising to fight "the greenhouse effect with the White House effect" (Boykoff and Boykoff 2007: 6; Peterson 1989). Surveys administered by Gallup in 1989, 1990, and 1991 that measured how much citizens were concerned about "global warming" found little partisan disagreement on the issue (see Egan and Mullin 2017, Fig. 3.2, Gallup polls, 1989–2016). In fact, politically attentive conservatives were more likely to believe scientists' warnings about global warming compared to liberals in the 1990s prior to when media coverage highlighted partisan disagreement and conflict on the issue (Tesler 2018). By 1997, however, concern about global warming had declined substantially among Republicans relative to Democrats. What was it that changed?

The roots of partisan divisions can be traced to the activities of interest groups and political elites that coalesced in the late 1980s and early 1990s, and increased media coverage of the issue surrounding political debate over whether the U.S. should ratify the Kyoto Protocol that committed signatories to reduce greenhouse gas emissions to mitigate climate change's effects (McCright and Dunlap 2003; Nisbet 2011; Oreskes and Conway 2010). The U.S. signed the agreement in 1997 but withdrew in 2001.

Beginning in the late 1980s, conservative donors began to fund think tanks and advocacy groups to dispute the growing scientific consensus about climate change's causes and impacts. The fossil fuel industry, for instance, financed a campaign designed to plant seeds of doubt among the public by emphasizing scientific uncertainty (McCright and Dunlap 2003; Oreskes and Conway 2010). The goal was to undermine support for regulatory policy action, in part by recruiting non-mainstream scientists to dispute the scientific consensus on the evidence for climate change and its anthropogenic causes (McCright and Dunlap 2000; Oreskes and Conway 2010).

Facing frequent accusations of political bias, mainstream journalists in the United States have attempted to demonstrate objectivity or "even-handed" news coverage by quoting or giving broadcast time to individuals on both sides of an issue. U.S. media coverage of the climate issue in the early and mid-1990s featured skeptics as often as it did mainstream climate scientists, despite the fact that the overwhelming majority of scientists agreed that human activity was responsible for accelerated climate change. Egan and Mullin (2017) explain, "As the public was learning about the problem in the 1990s, the mainstream media's adherence to the journalistic norm of balancing coverage between two sides of a dispute resulted in misrepresentations of climate change science, understating the scientific certainty about human contribution to the problem." The process contributed to a false perception among segments of the public about actual levels of scientific disagreement on the fundamentals of climate change (Boykoff and Boykoff 2004).

During this same period, Republican consultant Frank Luntz recommended in a strategy memo to fellow partisans that they emphasize "scientific uncertainty" surrounding climate change, and that the U.S. be described as bearing an unfair economic burden in the absence of international coordination to curb greenhouse gas emissions. This strategy was used effectively by conservative think tanks to defeat policy proposals aimed at reducing greenhouse gas emissions (McCright and Dunlap 2003; Jacques et al. 2008; Hmielowski et al. 2014). McCright and Dunlap (2003) analyzed hundreds of documents produced by over a dozen conservative think tanks between 1990 and 1997 and showed that climate skeptics used several tactics to plant seeds of doubt among the U.S. public, such as: (1) framing the supportive evidence as weak or nonexistent; (2) highlighting the benefits that might occur as a result of global warming; and, (3) emphasizing the economically harmful effects of policies to reduce greenhouse gas emissions. Conservative think tanks disseminated these messages in policy forums, public speeches, and press conferences in efforts to sway the public and policy-makers.

The debate over whether the U.S. should ratify the Kyoto Protocol in 1997 more than doubled newspaper coverage on climate change and increased television coverage of the issue by tenfold (Krosnick et al. 2000; Nisbet 2011). Most news coverage during this debate emphasized the scientific consensus regarding the reality of human impacts on climate change; however, media reports often included a discussion about "political conflict," and emphasized the actions of leading political figures, such as former Vice President Al Gore, in lobbying for the U.S. to ratify the agreement.

Media coverage on climate change, and partisan polarization, increased in the early 2000s. The release of the fictional movie, *The Day After Tomorrow*, in May 2004 that showed climate change causing a sudden freezing of cities. This movie generated 10 times more news coverage in the U.S. than the 2001 IPCC report emphasizing the dominant role of human activities in driving the observed warming trend (Leiserowitz 2005; Nisbet 2011). A defining feature of the climate disaster frame is that a looming catastrophe awaits unless action is taken immediately; however, an unintended consequence was that this portrayal, stretching the actual science, played into the hands of climate skeptics who labeled it "alarmism" that was motivated by a political and ideological agenda (Nisbet 2009).

The 2006 release of the film featuring former Vice President Al Gore, *An Inconvenient Truth,* led to a spike in the U.S. news media's coverage of climate change. Gore's message on climate change increased concern among fellow Democrats but reinforced disbelief among Republicans. The associated media coverage emphasizing political conflict is said to have helped to establish partisan divisions on climate change that persist (Jacques and Knox 2016).

Between 2010 and 2012, the percentage of the American public who reported believing in the existence of human caused climate change *declined* at a rate of about 5% per year according to two landline and cell phone surveys conducted on nationally representative samples (Krosnick and MacInnis 2012). This decline was most pronounced among conservative Republicans, an effect, in part, of media coverage of statements made by many of the Republican Party's leaders seeking the party's presidential nomination (Brulle et al. 2012; Carmichael et al. 2017; Krosnick and MacInnis 2012). In the US, higher levels of political interest and knowledge among conservatives and Republicans predicts greater skepticism about climate change, while higher knowledge and interest for liberals and Democrats predicts greater acceptance (Tesler 2018).

Media coverage on climate change varied widely across news outlets during this period. A content analysis of a sample of 642 articles on climate change appearing in *The New York Times*, *The Wall Street Journal (WSJ)*, *Washington Post*, and *USA Today* between 2006 and 2011 found significant differences in coverage comparing media frames across outlets (Feldman et al. 2017). The WSJ was more likely than the other news outlets both to mention the negative economic consequences that would result from efforts to tackle climate change and to accentuate political conflict. Another content analysis of news coverage surrounding climate change during this period found a dominance of "false-balanced" and "dismissive views" appearing in the conservative-leaning WSJ (Nisbet 2011). On cable news, Fox News was more likely than CNN or MSNBC to have guests making dismissive statements about climate science that challenged the "consensus view," and watching Fox

News was associated with a decreased belief that human-induced climate change is real and a decreased trust in scientists (Feldman et al. 2012; Hmielowski et al. 2014). Statements from leading Republicans seeking the presidency in 2012 likely reinforced the clear "partisan" division that had developed (Fig. 3.1).

Many surveys have found that the public ranks climate change as less important than a host of other economic, social, and political issues (Lorenzoni and Pidgeon 2006; Nisbet and Myers 2007; Motel 2014; Saad and Jones 2016). Yet, aggregate statistics mask a wide gap in the relative importance that partisans of different stripes assign to the issue. In 2016, for instance, liberal Democrats ranked climate change as the sixth most important problem facing the U.S. out of a list of 23, while moderate Republicans ranked it twenty-first and conservative Republicans ranked it very last (23rd) on the list (Bolsen and Shapiro 2018; Lehmann 2016). In a survey conducted by the Pew Research Center in 2016, only 38% of Americans thought the issue should be a "top priority" for lawmakers, ranking sixteenth out of a list of eighteen problems presented to respondents (Egan and Mullin 2017). The low sense of urgency the public attributes to dealing with climate change is partly due to the fact that it is an abstract problem that is difficult to directly perceive and experience, particularly in mid-latitude nations (van der Linden 2017; Weber 2010; Whitmarsh 2008).

The cumulative effect of climate science politicization, a steady diet of media coverage accentuating partisan conflict, and the development of increasingly partisan media has resulted in extreme levels of polarization on all aspects of the issue (Bolin and Hamilton 2018; Guber 2013; Guber 2017; Krosnick et al. 2000; Leiserowitz et al. 2018; Levendusky 2013; McCright and Dunlap 2011a; Palm et al. 2017; van der Linden 2017). Partisan polarization among the general public on climate change reflects analogous divisions among elite partisans in the U.S. (Brulle et al. 2012).

Rick Santorum	*"There is no such thing as global warming. .... It's just an excuse for more government control of your life, and I've never been for any scheme or even accepted the junk science behind the whole narrative." (Johnson and Somanader 2011).*
Mitt Romney	*"My view is that we don't know what's causing climate change on this planet. And the idea of spending trillions and trillions of dollars to try to reduce CO2 emissions is not the right course for us." (Anderson 2011).*
Newt Gingrich	*"I want to be clear: I don't think that we have conclusive proof of global warming. And I don't think we have conclusive proof that humans are at the center of it." (Jackson 2011).*
Rick Perry	*"The fact of the matter is the science is not settled on whether or not the climate change is being impacted by man to the point where we're going to put America's economics in jeopardy." (Lehmann 2011).*
Herman Cain	*"I don't believe ... global warming is real. Do we have climate change? Yes. Is it a crisis? No. ... Because the science, the real science, doesn't say that we have any major crisis or threat when it comes to climate change." (Dade 2011).*

Fig. 3.1 Statements on climate change by Republican candidates during the 2012 Primaries. (Krosnick and MacInnis 2012)

President Trump announced in June of 2017 that the United States would withdraw from the Paris climate accord – an agreement signed by 195 countries intended to reduce global greenhouse gas emissions. Under the agreement, the U.S. had pledged to cut emissions from 26 to 28% below 2005 levels by 2020 (Shear 2017). President Trump's decision to withdraw from the Paris agreement was praised by Republican Senate Majority Leader Mitch McConnell who stated, "I applaud President Trump and his administration for dealing yet another significant blow to the Obama administration's assault on domestic energy production and jobs" (Shear 2017). The President's decision to withdraw from the agreement and the supportive rhetoric from high-ranking Republican party elites likely reinforced and strengthened existing partisan polarization on climate change among the mass public (Tesler 2018).

In June 2019, White House officials barred an intelligence agency from submitting written testimony to the House Intelligence Committee warning of the "possibly catastrophic" consequences of climate change based on federal scientific findings (Eilperin et al. 2019). The report was prepared by the State Department's Bureau of Intelligence and Research and had originally stated, "Absent extensive mitigating factors or events, we see few plausible future scenarios where significant—possibly catastrophic—harm does not arise from the compounded effects of climate change" (Eilperin et al. 2019). A senior director from President Trump's National Security Council, William Happer, objected to the report's conclusions and to its use of the phrase "tipping points" when explaining how a certain level of warming could trigger devastating effects. Happer wrote, "[The report] includes lots of climate alarm propaganda that is not science at all… 'tipping points' is a propaganda slogan for the scientifically illiterate… they were a favorite of Al Gore's science adviser, James Hansen" (Eilperin et al. 2019). The report also documented how rising greenhouse gas emissions are acidifying the world's oceans and increasing the frequency and intensity of extreme weather events. President Trump responded to the report by stating: "I believe there's a change in weather, and I think it changes both ways… don't forget it used to be called global warming. That wasn't working. Then it was called climate change. Now it's actually called extreme weather because, with extreme weather, you can't miss" (Eilperin et al. 2019).

Partisanship now plays a primary role in the degree to which any scientific information about climate change disseminated by "experts" is perceived as credible and trustworthy (Daniels et al. 2017; Brewer and Ley 2013; Gauchat 2012; Kahan et al. 2011; Motta 2018). There is a reluctance to "trust the evidence" when it challenges prior beliefs (Kraft et al. 2015). Although there seems to have been an increase in the percentage of Republicans who believe that climate change is real and that a scientific consensus exists on the topic (Hamilton 2016), significant differences about the causes and therefore mitigation strategies persist. Recent surveys have found that Americans, particularly conservative Republicans, continue to underestimate the level of consensus among expert scientists regarding human-caused climate change, as well as the degree to which the general public believes it is happening (Jost 2018; Mildenberger and Tingley 2017). A key hurdle facing science communicators seeking policy action is thus how to generate greater awareness and engagement among a highly partisan audience in the U.S., (Akerlof et al. 2016; Druckman 2013; Druckman 2015; Lupia 2013; Priest 2016; Sussman and Byron 2013).

3.2 Demographic, Social, and Personal Predictors of Climate Change Beliefs in the US

Although party identification and ideology are the strongest predictors of climate change beliefs and support for action in the United States, other demographic characteristics, cultural worldviews, and personal experiences are also associated with these attitudes and behaviors. First, several studies have shown that women and persons of color express higher levels of belief in and concern about climate change (Brody et al. 2008; Egan and Mullin 2012, 2017; Godfrey and Torres 2016; Leiserowitz 2006; McCright and Dunlap 2011b), which is consistent with the literature on race, gender and risk aversion (Slovic 1999). This stable finding has been referred to as the "white male effect" (Finucane et al. 2000), which indicates that white males, especially those who identify as political conservatives, are generally less concerned about climate change as well as other environmental risks (McCright and Dunlap 2011b). A number of surveys have shown that females and non-whites are more worried about climate change than are white males, perhaps due to the increased vulnerability to negative environmental impacts and hardships for people with fewer economic and political resources (Bord and O'Connor 1997; Leiserowitz 2006; Schuldt and Pearson 2016; van der Linden 2015; West and Orr 2007).

A number of other demographic characteristics are less consistently associated with climate change opinions, including age, education, and income. For instance, although some studies have found that higher levels of education predict greater awareness of and concern about climate change (Hornsey et al. 2016; Lee et al. 2015; van der Linden 2015), other research has reported that higher levels of knowledge among partisans in the U.S. *increases* the gap between Democrats and Republicans over the fundamental science surrounding climate change (Bolsen et al. 2015; Hamilton 2011; Kahan et al. 2011; Malka et al. 2009).

There are equally inconsistent results for how age influences concern about climate change. While some studies show a negative correlation between being older and climate risk perception (Heath and Gifford 2006; Hornsey et al. 2016; Kellstedt et al. 2008; Malka et al. 2009; Milfont 2012), others show no relationship (Akerlof et al. 2013; O'Connor et al. 1999; Sundblad et al. 2007), or even a positive relationship (Slimak and Dietz 2006). While income might be assumed to decrease individuals' vulnerability to the impacts of climate change, there are mixed findings regarding the association between income and perceived risks (Akerlof et al. 2013; Hornsey et al. 2016; Smith and Leiserowitz 2012). There is thus some evidence that correlates socio-demographic factors with climate change beliefs, but these variables are often included as control variables in models with little theorizing about their conceptual relevance (Dietz et al. 1998; van der Linden 2017).

Cultural worldviews, particularly orientations to individualism and hierarchy, have also been posited to affect beliefs about climate change (Kahan et al. 2012; Kahan 2013; Leiserowitz 2006). "Individualism" is seen to predict skepticism about environmental risks and opposition to policy that might bring about further regulation or restrictions on personal freedom (Douglas and Wildavsky 1982). Those who value "hierarchy" place more trust in perceived experts, particularly those whose worldviews they share (Kahan et al. 2011). According to "cultural cognition theory (CCT)," people form perceptions of risk that "connect them to others who share their cultural values… [and they form] 'worldviews' or preferences for how to organize society or other collective enterprise … along two orthogonal dimensions" (Kahan 2015). On the first dimension, people who value individualism prefer a social ordering that places the highest priority on the protection of individual liberty, and views individuals as responsible for collective outcomes. At the other end of this same dimension, "communitarians" place relatively less value on liberty and are more willing to place restrictions on individual liberties to promote the collective good. Communitarians also see government as primarily responsible for collective outcomes. On the second dimension, CCT posits that individuals who value hierarchy prefer a social ordering that is "pervaded with rankings that tie authority to social roles" (Kahan 2015). At the opposite end of the second dimension, "egalitarians" reject social stratification, view nature as fragile and the environment as needing protection. The basic premise of CCT is that people selectively credit or dismiss evidence about societal risks based on whether it supports or contradicts their own values (Kahan et al. 2011; Kahan 2015).

Both correlational and experimental evidence have demonstrated that these worldviews shape beliefs about climate change and response to information that either supports or contradicts their existing cultural identities and associated beliefs (Kahan 2013). Moreover, as those who possess different partisan or ideological identities become more factually knowledgeable about climate change, they become increasingly polarized along cultural and ideological lines (Bolsen et al. 2015; Kahan 2015; Hamilton 2011). These findings indicate that attempting to increase the scientific literacy of the public about the "facts" surrounding climate change may do little to reduce polarization. In fact, those individuals who possess higher levels of "cognitive expertise" (e.g., greater numeracy, increased knowledge, etc.) are better equipped to engage in "identity protective cognition," a form of "directional motivated reasoning." This is because they are more skilled at understanding the degree to which scientific or political information reflects positively or negatively on their own group identities and cultural worldviews (Bolsen et al. 2015; Kahan et al. 2012; Kahan 2015).

A person's degree of religiosity has also been associated with belief in and concern about climate change, especially among certain Christian denominations (Arbuckle and Konisky 2015; Egan and Mullin 2012; McCright and Dunlap 2011a). Religiosity, however, seems to have a limited impact, with some studies finding only a small and inconsistent effect (Hamilton and Keim 2009; Smith and Leiserowitz 2012).

People often use their own experiences to make judgments when they lack more detailed information. The local temperature that people experience daily, for instance, is one of the most readily accessible experiences that has been shown to shape Americans' beliefs about climate change (Egan and Mullin 2012; Hamilton and Stampone 2013; Schuldt and Roh 2014; Weber and Stern 2011). Laboratory experiments and surveys consistently show a positive correlation between exposure to hotter temperatures, for example in the summer months, and increased belief in global warming (Joireman et al. 2010; Shao et al. 2014; Zaval et al. 2014). Although short-term weather-related events can produce meaningful shifts in people's beliefs about climate change, these effects appear to be relatively short-lived and can be counteracted by prompting people to consider longer-term weather trends (Druckman 2015; Druckman and Shafranek 2017; Konisky et al. 2016). In addition, although there is a robust relationship between climate change risk perceptions and experiences with extreme weather events such as hurricanes, flooding, heat waves, and droughts (Akerlof et al. 2013; Capstick and Pidgeon 2014; Howe et al. 2015; Krosnick et al. 2006; Myers et al. 2012; Reser et al. 2014; van der Linden 2015), the magnitude of the association is small in comparison with the influence of party identification and political ideology (Zia and Todd 2010).

3.3 Strategic Messaging to Influence Climate Change Beliefs

A great deal of survey and laboratory research has shown the overwhelming influence of political partisanship on climate change opinion in the United States and has also delved into some of the underlying cognitive processes involved in decision-making. As in so many other realms of persuasive communication, the way the message is framed, the credibility of the source, the fears, hopes and other emotions connected with accepting the message and the behavior changes that it suggests, and the prior beliefs and values of the audience all play a significant role (Anshelm and Hultman 2014; Druckman and Bolsen 2011; Flottum 2017; Nisbet 2009; Hart 2011; Myers et al. 2012; Nerlich et al. 2010). Most survey questions that have tapped the opinions of US residents on climate change over the past two decades have focused on four metrics: (a) individuals' awareness and knowledge about climate change, (b) the extent to which climate change is occurring, (c) personal concern and the relative importance of the issue, and (d) support for specific types of policy (Egan and Mullin 2017). Several key findings have emerged from research that studies how strategically framed messages influence climate change perceptions and decisions (for a detailed review, see Bolsen and Shapiro 2017; Moser 2010; Moser and Dilling 2011).

First, strategically framed messages can shift individuals' opinions and reported willingness to take action on climate change (Feldman and Hart 2018; Hart 2011; Lewandowsky et al. 2013; McCright et al. 2016; Myers et al. 2012; van der Linden et al. 2019). A communicator, for instance, might highlight the economic, environmental, public health, and/or moral effects associated with climate change, thereby providing an "interpretive storyline that set[s] a specific train of thought in motion, communicating why an issue might be a problem, who or what might be responsible for it, and what should be done about it" (Nisbet 2009). Messages that emphasize climate change's local impacts, effects on public health, the degree to which those in a person's social group are concerned about the issue, degree of scientific consensus, or high levels of public concern shift beliefs in and concern about climate change (Bolsen et al. 2014a, b; Goldberg et al. 2019; Levine and Kline 2017; O'Neill 2013; O'Neill and Nicholson-Cole 2009; Leiserowitz 2006; Lewandowsky et al. 2013; Scannell and Gifford 2013; Schweizer et al. 2013).

Several studies explore how visual images that are common in media reports on climate change, such as photographs highlighting political conflict or geographically remote locations, affect individuals' perceptions about the threat posed by climate change and concern over its effects (O'Neill 2013; O'Neill and Nicholson-Cole 2009). For instance, highlighting the local impacts of climate change through visual images showing flooding that would occur as a result of sea-level rise in coastal communities, can increase public concern and support for actions to address the problem (Bolsen et al. 2018). However, response to otherwise credible and persuasive information can be undermined when presented in competition with rhetoric that politicizes climate science, for instance by those seeking to cast doubt on the existence of a scientific consensus (Bolsen and Druckman 2018; Druckman 2017; Jacques and Knox 2016; McCright and Dunlap 2011a; McCright et al. 2016; van der Linden et al. 2017). Nonetheless, visual frames that accentuate the local effects of climate change—for instance, by highlighting local flooding that would occur in coastal communities in the U.S. as a result of sea-level rise—can be an effective vehicle for promoting greater belief in the fundamentals of climate science. They may also be effective in combating climate science's politicization (Bolsen et al. 2019a; Hart and Feldman 2016; Leiserowitz 2007; Retchless 2018; Spence and Pidgeon 2010; Walsh 2015).

Other research has tested the impact of providing information about the degree to which a scientific consensus exists on climate change on individuals related beliefs and willingness to support climate action. A number of studies have showed that the perception that a scientific consensus exists is an important "gateway belief" predictive of whether or not a person will report believing that human-caused climate change is real and support for climate policy action (Lewandowsky et al. 2013; van der Linden et al. 2015a, b; van der Linden 2017). Lewandowsky et al. (2013) found that exposure to a frame stating that 97% of climate scientists believe in human-caused climate change increases individuals' perception that a scientific consensus exists and influences other fundamental beliefs about climate change and ultimately increasing support for action.

Yet the circumstances and contexts in which providing individuals with consensus scientific information as a way to shift beliefs toward the consensus position continues to be a topic of debate among scholars (Dixon and Hubner 2018; Kahan 2016; van der Linden 2016; van der Linden et al. 2017). On one hand, there is evidence that messages highlighting levels of expert scientific consensus on climate change can lead to greater belief in this scientific consensus (Deryugina and Shurchkov 2016; Ranney and Clark 2016; van der Linden 2016); on the other hand, consensus climate messages are *not always effective in changing other beliefs or behaviors, particularly for* those who possess strong contradictory prior beliefs or who are motivated to process information in a manner that allows them to protect existing worldviews or identity-defining groups (Hart and Nisbet 2012; Kahan et al. 2011; Kahan et al. 2012; Kahan 2015; Ma et al. 2019; Mildenberger et al. 2019).

A second finding is that prior beliefs, group identities, and cultural worldviews often moderate response to targeted communication efforts on climate change. This occurs because of a phenomenon called "motivated reasoning" (Bolsen and Palm 2019). There are two types of motivations that scholars have studied in research on climate opinion formation: an accuracy motivation and a directional motivation. When individuals are orientated toward an accuracy motivation, they evaluate information in an "even-handed" fashion in an attempt to arrive at the "best" or "most correct" belief outcome (Druckman 2012; Kunda 1990; Taber and Lodge 2006). On the other hand, individuals who seek to protect their existing beliefs or partisan identity may process information in a way that upholds their existing beliefs or group commitments. This is referred to as "directional motivated reasoning." A directional motivation causes individuals to seek out information that confirms their existing beliefs or identities (a confirmation bias), dismiss or counter-argue against contradictory arguments and evidence (a disconfirmation bias), and perceive belief- or identity-confirming arguments and evidence as stronger than disconfirming information (a prior attitude effect).

In the specific case of motivated reasoning affecting climate change beliefs, Democrats and Republicans can be provided with identically framed messages but may respond quite differently as a result of the underlying information processing motivation "biasing" the way in which the messages are evaluated (Cook and Lewandowsky 2016). Partisan belief polarization can occur *both* as a result of *directional* motivations *or accuracy motivations* (Druckman and McGrath 2019). For example, differences in assessments of what is considered "accurate information" may actually stem from differences in *prior beliefs about the credibility of any message* based, for instance, on perceptions about the trustworthiness of its source (Landrum et al. 2015; Landrum et al. 2017; Lupia and McCubbins 1998).

The actual operation of motivated reasoning is complex, however. While the most prominent explanation for partisan polarization is directional motivated reasoning, it could equally be that prior beliefs are causing people to find some information more or less credible (Druckman and McGrath 2019). To test this hypothesis, Bolsen and Druckman (2018) provided identical consensus information about climate change to Republicans and Democrats. Low-knowledge Democrats and Republicans, and high knowledge Democrats, showed an increase in their belief in human-caused climate change. High knowledge Republicans, conversely, were not influenced by the message. In such a case, it is impossible to know, however, whether high knowledge Republicans were engaging in motivated reasoning to defend their prior belief (a "directional" goal) or if they were pursuing an accuracy motivation but were skeptical about the validity or credibility of the scientific information that was provided. This distinction in *process* is crucial because the success of any communication effort related to climate change depends on motivation: if individuals are striving for accuracy, then establishing the credibility of the communicator of a message is of paramount importance. Such efforts will be futile however, if directional goals are driving polarization (Druckman and McGrath 2019).

Other scholarship has found that when individuals are presented information that contradict their prior beliefs, they move in the opposite direction of the intended message. This has been labeled as "boomerang," "backlash," or "backfire" effects (Byrne and Hart 2009; Chapman and Lickel 2016; Cook and Lewandowsky 2016; Gromet et al. 2013; Hart and Nisbet 2012; Kahan et al. 2012; Ma et al. 2019; Nisbet et al. 2015; Nyhan and Reifler 2010; Zhou 2016). For instance, Zhou (2016) found that messages promoting government action on climate change that emphasized economic, national security, moral justice, or natural disaster implications had no effect on Republicans, or resulted in backfire effects when the message was sponsored by a Democratic congressman (as opposed to a Republican congressman). The results suggest that disconfirmation biases led respondents to generate additional "counter-arguments" that caused them to update their opinion in the opposite direction of the message (also see, Nyhan and Reifler 2010). In a similar line of research, Kahan et al. (2011) conducted a study to test the hypothesis that cultural cognition driven by motivated reasoning explains the failure of expert consensus to dispel public controversy over the issues of climate change, nuclear waste storage, and gun rights. They conducted a nationally representative survey experiment in the U.S. and varied attributions of a "low risk" or "high risk" frame on each issue to one of three featured (randomized) scientists. In strong support of motivated reasoning as driving information processing and opinion formation in this instance, participants' evaluations of whether the scientist was an expert depended on the fit between the position the scientist was taking on a given issue and the position that dominated among the subject's own cultural groups.

Pro-climate action messages also have been found to backfire due to a process called "psychological reactance" (Brehm 1966; Nisbet et al. 2015). Reactance occurs when people's choices are reduced, because they do not like to be told "what to do, how to act, or what to believe" (Ma et al. 2019). Messages that threaten people's political views are particularly prone to generating reactance (Meirick and Nisbett 2011; Weeks and Garrett 2014). For instance, several studies have explored how messages highlighting the influence of climate change on increased frequency of extreme weather events and natural hazards influence the belief in the existence of climate change (Dixon et al. 2019; Ma et al. 2019; Mildenberger et al. 2019). Drawing attention to these events may increase the personal salience of the issue and focus individuals on the impacts of climate change on their own homes and communities (Weber 2010). While some have proposed using extreme weather frames to raise awareness about climate change and persuade skeptical audiences (Cody et al. 2017; Rudman et al. 2013), several studies that have tested this approach have found that linking climate change to such events leads to message resistance among skeptical audiences. For instance, Mildenberger et al. (2019) conducted a large-scale survey experiment in the San Francisco Bay Area to explore how providing residents with personally-relevant risk messages influenced climate risk perceptions in the context of sea-level rise. In the treatment, the authors showed respondents one of two maps. The first was a map of their home zip (postal) code showing the effects of sea-level rise of .98 meters (3.2 feet). The second was the same map but with the sea-level rise flooding coupled with a 100-year storm scenario. They found that, relative to those assigned to a control group who did not receive any information, exposure to a map showing flooding at the zip-code level significantly *decreased* residents' belief that sea-level rise would threaten the future of their communities and had no impact on other general measures of concern about the risk of future sea-level rise.

A third finding is that although polarization, motivated reasoning, and low levels of public engagement pose serious challenges to those seeking to promote action, several best practices in climate change communication have been identified. Among them are describing climate change as a personal risk, using social group norms to convince skeptics, and appealing to the values that the audience shares. Frames that translate the effects of climate change into personal experiences can shape risk perceptions across individuals who hold different identities and worldviews (Egan and Mullin 2012; van der Linden 2017). This turns the risks associated with climate change into concrete and relatable personal experiences that can exert a powerful impact on beliefs and support for action.

Perception of social consensus on the issue of climate change, particularly within ones' close networks, also influences people to align their beliefs with what they perceive to be the majority point-of-view (Goldberg et al. 2019; Jost 2018; van der Linden 2015). It should be noted, however, that this effect is attenuated for Americans, since they tend to underestimate the actual level of social consensus about human-induced climate change (Mildenberger and Tingley 2017). Nonetheless, empirical research has identified the power of social consensus messages. In a recent experiment that varied sources associated with a ballot proposal, Republicans were more likely to support a climate proposal when it was endorsed by Republican party elites, and perceptions that other Republicans also supported the initiative was a key mediating belief (Ehret et al. 2018). In another recent study, a message sponsored by Republican senators led Republican respondents to reject misinformation about climate change, increase their belief in the existence of a scientific consensus, and increase their belief that it is human-caused (Benegal and Scruggs 2018). In these cases, the partisan membership of the person delivering the message plays a powerful role in the degree to which an appeal persuades any audience (Benegal and Scruggs 2018; Ehret et al. 2018; Kousser and Tranter 2018; Li et al. 2016; Maibach et al. 2015; Van Boven et al. 2018).

Messages that are framed in ways that appeal to the values of conservatives and Republicans, such as by highlighting free-market-oriented solutions to climate change, appear to be another mechanism for shifting the views of skeptics (Bain et al. 2012; Dixon et al. 2017; Feinberg and Willer 2013). For instance, Wolsko et al. (2016) found that conservatives who were exposed to a "binding morality frame" (protecting one's homeland) as opposed to an individualizing morality frame (care for the environment) or a control group expressed greater belief in and concern about climate change (but see, Severson and Coleman 2015). Kahan et al. (2015) found that highlighting free-market geoengineering solutions to those who value individualism, and who are often climate skeptics, leads them to see evidence in support of climate change as more credible (also see, Campbell and Kay 2014). In addition, messages that affirm the identity or value of an individual can result in any new information being evaluated in a "non-threatening" manner that allows belief updating to occur without disconfirmation biases (Druckman and McGrath 2019).

However, it isn't always clear which frames will resonate with particular audiences. Some have argued that "rather than continually testing the impact of one frame after another, the literature would benefit from exploring the conditions that lead to distinct types of motivations, and then investigating which types of messages resonate in light of motivations and particular prior beliefs, values and identities" (Druckman and McGrath 2019).

What is incontrovertible is the need for credibility among those who step forward to reach diverse audiences who may be skeptical about climate change. Only frames from credible sources are likely to be impactful in shaping opinions (Druckman 2001; Landrum et al. 2015, 2017), especially in politicized environments where there can be suspicion about the true motives and expertise of a science communicator and whether or not there are shared common interests (Kahan et al. 2011; Lupia 2013). One solution might be to convene scientists and communicators who hold varied sets of values and ideological leanings with the goal of finding and communicating consensus about fundamental science. Given the current partisan divide on the issue of climate change, Hoffman (2015) argues that "it is only logical that more climate brokers must emerge from the political Right… [C]limate change must be presented by bona fide conservatives and conservative organizations that see the solutions to climate change as consistent with a conservative ideology." In other words, conservatives and Republicans are more likely to be persuaded by information provided by credible leaders who they trust (i.e. fellow partisans and ideologies). Nisbet (2014) explains, "… once community members from differing political backgrounds join together to achieve a broadly inspiring goal like protecting people and a local way of life, then the networks of trust and collaborations formed can be used to move this diverse segment toward cooperation in pursuit of national policy goals."

The perceived credibility of any information source is crucial (O'Keefe 2012). One recent study found that the impact of identical arguments about the environmental or national security effects that will happen due to climate change are conditioned by whether the information is attributed to Democratic Party leaders, Republican Party leaders, military experts, or climate scientists. Republican respondents who received messages sponsored by an in-party source (i.e., Republican Party leaders) displayed more concern about climate change and belief in the scientific consensus relative to an identical appeal linked with an out-group (i.e., Democratic Party leaders), or especially relative to climate scientists – who generated a strong negative reaction among Republican respondents (Bolsen et al. 2019b).

Messages that emphasize "positive solutions" for the nation's future have also been suggested as a best practice for communicating climate information. Climate solutions are often framed as immediate losses for society that involve reductions in energy consumption and travel, and requiring higher taxes (van der Linden et al. 2015a, b). Communicators must work to shift the policy conversation away from such "losses" and toward the positive gains of immediate policy action – e.g., both economic and environmental reasons. For example, Kahan et al. (2011) found that when hierarchical-individualists were exposed to a frame that a solution to climate change is not to regulate emissions but to invest in nuclear power, their support for policy action on climate change increased. This is consistent with research which has found that Republican and conservative individuals often say that they do not believe in climate change because of an aversion of the policy solutions that would follow, such as new laws and regulations that would restrict freedom and personal choice (Campbell and Kay 2014).

3.4 Summary

Simply providing information about the existence of climate change and its likely consequences is insufficient to change minds. More effective strategies will be needed. Based on the findings of a large body of prior research, we expected a more intense response to sea-level rise when the message provided was local and the immediate threat was visible. With these expectations, we selected our study population and designed the survey instrument.

References

Akerlof K, Maibach E, Fitzgerald D, Cedeno A, Neuman A (2013) Do people personally experience global warming, and if so how, and does it matter? Glob Environ Chang 23(1):81–91. https://doi.org/10.1016/j.gloenvcha.2012.07.006

Akerlof K, Rowan K, La Porte T, Batten B, Ernst H, Sklarew D (2016) Risky business: engaging the public on sea-level rise and inundation. Environ Sci Pol 66:314–332. https://doi.org/10.1016/j.envsci.2016.07.002

Amadeo K (2019) Hurricane Irma facts, damage, and costs. In: The Balance. Available at: https://www.thebalance.com/hurricane-irma-facts-timeline-damage-costs-4150395. Accessed 12 Mar 2019

Anderson D (2011) Mitt Romney talks climate change in Dover, New Hampshire. In: Newhampshireprimary.blogspot.com. Available at: http://newhampshireprimary.blogspot.com/2011/08/mitt-romney-talks-climate-change-in.html. Accessed 2 Feb 2018

Anshelm J, Hultman M (2014) Discourses of global climate change: apocalyptic framing and political antagonisms. Routledge, London

Arbuckle M, Konisky D (2015) The role of religion in environmental attitudes. Soc Sci Q 96(5):1244–1263. https://doi.org/10.1111/ssqu.12213

Bain P, Hornsey M, Bongiorno R, Jeffries C (2012) Promoting pro-environmental action in climate change deniers. Nat Clim Chang 2(8):600–603. https://doi.org/10.1038/nclimate1532

Benegal SD, Scruggs LA (2018) Correcting misinformation about climate change: the impact of partisanship in an experimental setting. Clim Chang 148(1–2):61–80. https://doi.org/10.1007/s10584-018-2192-4

Bolin J, Hamilton L (2018) The news you choose: news media preferences amplify views on climate change. Environ Polit 27(3):455–476. https://doi.org/10.1080/09644016.2018.1423909

Bolsen T, Druckman J (2018) Do partisanship and politicization undermine the impact of a scientific consensus message about climate change? Group Process Intergroup Relat 21(3):389–402. https://doi.org/10.1177/1368430217737855

Bolsen T, Palm R (2019) Motivated reasoning and political decision making. In: Oxford encyclopedia of political decision making. Oxford University Press, Oxford. https://doi.org/10.1093/acrefore/9780190228637.013.923

Bolsen T, Shapiro MA (2017) Strategic framing and persuasive messaging to influence climate change perceptions and decisions. In: Oxford research encyclopedia of climate science. https://doi.org/10.1093/acrefore/9780190228620.013.385

Bolsen T, Shapiro M (2018) The US news media, polarization on climate change, and pathways to effective communication. Environ Commun 12(2):149–163. https://doi.org/10.1080/17524032.2017.1397039

Bolsen T, Druckman J, Cook F (2014a) How frames can stunt support for scientific adaptations: politicization and the status-quo bias. Public Opin Q 78(1):1–26. https://doi.org/10.1093/poq/nft044

Bolsen T, Leeper T, Shapiro M (2014b) Doing what others do: norms, science, and collective action on global warming. Am Politics Res 42(1):17–30. https://doi.org/10.1177/1532673X13484173

Bolsen T, Druckman J, Cook F (2015) Citizens', scientists', and policy advisors' beliefs about global warming. Ann Am Acad Pol Soc Sci 658(1):271–295. https://doi.org/10.1177/0002716214558393

Bolsen T, Kingsland J, Palm R (2018) The impact of frames highlighting coastal flooding in the USA on climate change beliefs. Clim Chang 147(1–2):359–368. https://doi.org/10.1007/s10584-018-2143-0

Bolsen T, Palm R, Kingsland J (2019a) Counteracting climate science politicization with effective frames and imagery. Sci Commun 41(2):147–171

Bolsen T, Palm R, Kingsland J (2019b) The impact of message source on the effectiveness of communications about climate change. Unpublished manuscript. Georgia State University, Atlanta

Bord R, O'Connor R (1997) The gender gap in environmental attitudes: the case of perceived vulnerability to risk. Soc Sci Q:830–840

Boykoff M, Boykoff J (2004) Balance as bias: global warming and the U.S. prestige press. Glob Environ Chang 14(2):125–136. https://doi.org/10.1016/j.gloenvcha.2003.10.001

Boykoff M, Boykoff J (2007) Climate change and journalistic norms: a case-study of U.S. mass-media coverage. Geoforum 38(6):1190–1204. https://doi.org/10.1016/j.geoforum.2007.01.008

Brehm J (1966) A theory of psychological reactance. Academic, New York

Brewer P, Ley B (2013) Whose science do you believe? Explaining trust in sources of scientific information about the environment. Sci Commun 35(1):115–137. https://doi.org/10.1177/1075547012441691

Brody S, Zahran S, Vedlitz A, Grover H (2008) Examining the relationship between physical vulnerability and public perceptions of global climate change in the United States. Environ Behav 40(1):72–95. https://doi.org/10.1177/0013916506298800

Brulle R, Carmichael J, Jenkins J (2012) Shifting public opinion on climate change: an empirical assessment of factors influencing concern over climate change in the US, 2002–2010. Clim Chang 114(2):169–188. https://doi.org/10.1007/s10584-012-0403-y

Byrne S, Hart P (2009) The boomerang effect: a synthesis of findings and a preliminary theoretical framework. Ann Int Commun Assoc 33(1):3–37. https://doi.org/10.1080/23808985.2009.11679083

Campbell T, Kay A (2014) Solution aversion: on the relation between ideology and motivated disbelief. J Pers Soc Psychol 107(5):809–824. https://doi.org/10.1037/a0037963

Capstick S, Pidgeon N (2014) Public perception of cold weather events as evidence for and against climate change. Clim Chang 122(4):695–708. https://doi.org/10.1007/s10584-013-1003-1

Carmichael J, Brulle R, Huxster J (2017) The great divide: understanding the role of media and other drivers of the partisan divide in public concern over climate change in the USA, 2001–2014. Clim Chang 141(4):599–612. https://doi.org/10.1007/s10584-017-1908-1

Chapman A, Lickel B (2016) Climate change and disasters: how framing affects justifications for giving or withholding aid to disaster victims. Soc Psychol Personal Sci 7(1):13–20. https://doi.org/10.1007/s10584-017-1908-1

Cody E, Stephens J, Bagrow J, Dodds P, Danforth C (2017) Transitions in climate and energy discourse between hurricanes Katrina and Sandy. J Environ Stud Sci 7(1):87–101. https://doi.org/10.1007/s13412-016-0391-8

Cook J, Lewandowsky S (2016) Rational irrationality: modeling climate change belief polarization using Bayesian networks. Top Cogn Sci 8(1):160–179. https://doi.org/10.1111/tops.12186

Cook J, Oreskes N, Doran P, Anderegg W, Verheggen B, Maibach E et al (2016) Consensus on consensus: a synthesis of consensus estimates on human caused global warming. Environ Res Lett 11(4). https://doi.org/10.1088/1748-9326/11/4/048002

Dade C (2011) In their own words: GOP candidates and science. In: National Public Radio. Available at: http://www.npr.org/2011/09/07/140071973/in-their-own-words-gop-candidates-andscience. Accessed 24 June 2018

Daniels T, Shafer M, Shaikh H, Harrington C, Wagstaff L, Bieniek-Tobasco A, McCormick S (2017, April) Short-term effects of climate change communication: how political party affiliation affects receptivity to environmental messages. Poster presented at GW Milken institute school of public health poster presentations, George Washington University

Deryugina T, Shurchkov O (2016) The effect of information provision on public consensus about climate change. PLoS One 11(4):e0151469. https://doi.org/10.1371/journal.pone.0151469

Dietz T, Stern P, Guagnano G (1998) Social structural and social psychological bases of environmental concern. Environ Behav 30(4):450–471. https://doi.org/10.1177/001391659803000402

Dixon G, Hubner A (2018) Neutralizing the effect of political worldviews by communicating scientific agreement: a thought-listing study. Sci Commun 40(3):393–415. https://doi.org/10.1177/1075547018769907

Dixon G, Hmielowski J, Ma Y (2017) Improving climate change acceptance among US conservatives through value-based message targeting. Sci Commun 39(4):520–534. https://doi.org/10.1177/1075547017715473

Dixon G, Bullock O, Adams D (2019) Unintended effects of emphasizing the role of climate change in recent natural disasters. Environ Commun 13(2):135–143. https://doi.org/10.1080/17524032.2018.1546202

Douglas M, Wildavsky A (1982) How can we know the risks we face? Why risk selection is a social process 1. Risk Anal 2(2):49–58. https://doi.org/10.1111/j.1539-6924.1982.tb01365.x

Druckman J (2001) On the limits of framing effects: who can frame? J Polit 63(4):1041–1066. https://doi.org/10.1111/0022-3816.00100

Druckman J (2012) The politics of motivation. Crit Rev 24(2):199–216. https://doi.org/10.1080/08913811.2012.711022

Druckman J (2013) Public opinion: stunted policy support. Nat Clim Chang 3(7):617. https://doi.org/10.1038/nclimate1939

Druckman J (2015) Communicating policy-relevant science. Polit Sci Polit 48(1):58–69. https://doi.org/10.1017/S1049096515000438

Druckman J (2017) The crisis of politicization within and beyond science. Nat Hum Behav 1(9):615. https://doi.org/10.1038/s41562-017-0183-5

Druckman J, Bolsen T (2011) Framing, motivated reasoning, and opinions about emergent technologies. J Commun 61(4):659–688. https://doi.org/10.1111/j.1460-2466.2011.01562.x

Druckman J, McGrath M (2019) The evidence for motivated reasoning in climate change preference formation. Nat Clim Chang 9:111–119. https://doi.org/10.1038/s41558-018-0360-1

Druckman J, Shafranek R (2017) The conditional nature of the local warming effect. Weather Clim Soc 9(1):15–26. https://doi.org/10.1175/WCAS-D-16-0012.1

Dunlap R, McCright A, Yarosh J (2016) The political divide on climate change: partisan polarization widens in the US. Environ Sci Policy Sustain Dev 58(5):4–23. https://doi.org/10.1080/00139157.2016.1208995

Egan P, Mullin M (2012) Turning personal experience into political attitudes: the effect of local weather on Americans' perceptions about global warming. J Polit 74(3):796–809. https://doi.org/10.1017/S0022381612000448

Egan P, Mullin M (2017) Climate change: US public opinion. Annu Rev Polit Sci 20(1):209–227. https://doi.org/10.1146/annurev-polisci-051215-022857

Ehret P, Van Boven L, Sherman D (2018) Partisan barriers to bipartisanship: understanding climate policy polarization. Soc Psychol Personal Sci 9(3):308–318. https://doi.org/10.1177/1948550618758709

Eilperin J, Dawsey J, Dennis B (2019) White House blocked intelligence agency's written testimony calling climate change 'possibly catastrophic'. In: The Washington Post. Available at: https://www.washingtonpost.com/climate-environment/2019/06/08/white-house-blocked-intelligence-aides-written-testimony-saying-human-caused-climate-change-could-be-possibly-catastrophic/?utm_term=.4188340b6927. Accessed 9 June 2019

Feinberg M, Willer R (2013) The moral roots of environmental attitudes. Psychol Sci 24(1):56–62. https://doi.org/10.1177/0956797612449177

Feldman L, Hart P (2018) Is there any hope? How climate change news imagery and text influence audience emotions and support for climate mitigation policies. Risk Anal 38(3):585–602. https://doi.org/10.1111/risa.12868

Feldman L, Maibach E, Roser-Renouf C, Leiserowitz A (2012) Climate on cable: the nature and impact of global warming coverage on Fox News, CNN, and MSNBC. Int J Press/Polit 17(1):3–31. https://doi.org/10.1177/1940161211425410

Feldman L, Hart PS, Milosevic T (2017) Polarizing news? Representations of threat and efficacy in leading US newspapers' coverage of climate change. Public Underst Sci 26(4):481–497. https://doi.org/10.1177/1075547013520239

Finucane M, Slovic P, Mertz C, Flynn J, Satterfield T (2000) Gender, race, and perceived risk: the 'white male' effect. Health Risk Soc 2(2):159–172. https://doi.org/10.1080/713670162

Flottum K (ed) (2017) The role of language in the climate change debate. Routledge, London

Gauchat G (2012) Politicization of science in the public sphere: a study of public trust in the United States, 1974 to 2010. Am Sociol Rev 77(2):167–187. https://doi.org/10.1177/0003122412438225

Godfrey P, Torres D (eds) (2016) Systemic crises of global climate change: intersections of race, class and gender. Routledge, London

Goldberg M, van der Linden S, Leiserowitz A, Maibach E (2019) Perceived social consensus can reduce ideological biases on climate change. Environ Behav. https://doi.org/10.1177/0013916519853302

Gromet D, Kunreuther H, Larrick, R (2013) Political ideology affects energy-efficiency attitudes and choices. Proc Natl Acad Sci 110 (23): 9314–9319. Clark W (ed). https://doi.org/10.1073/pnas.1218453110

Guber D (2013) A cooling climate for change? Party polarization and the politics of global warming. Am Behav Sci 57(1):93–115. https://doi.org/10.1177/0002764212463361

Guber D (2017) Partisan cueing and polarization in public opinion about climate change. In: Oxford research encyclopedia of climate science. https://doi.org/10.1093/acrefore/9780190228620.013.306

Hamilton L (2011) Education, politics and opinions about climate change evidence for interaction effects. Clim Chang 104(2):231–242. https://doi.org/10.1007/s10584-010-9957-8

Hamilton L (2016) Public awareness of the scientific consensus on climate. SAGE Open 6(4). https://doi.org/10.1177/2158244016676296

Hamilton L, Keim B (2009) Regional variation in perceptions about climate change. Int J Climatol 29(15):2348–2352. https://doi.org/10.1002/joc.1930

Hamilton L, Stampone M (2013) Blowin' in the wind: short-term weather and belief in anthropogenic climate change. Weather Clim Soc 5(2):112–119. https://doi.org/10.1175/WCAS-D-12-00048.1

Hart P (2011) One or many? The influence of episodic and thematic climate change frames on policy preferences and individual behavior change. Sci Commun 33(1):28–51. https://doi.org/10.1177/1075547010366400

Hart P, Feldman L (2016) The impact of climate change–related imagery and text on public opinion and behavior change. Sci Commun 38(4):415–441. https://doi.org/10.1177/1075547016655357

Hart P, Nisbet E (2012) Boomerang effects in science communication: how motivated reasoning and identity cues amplify opinion polarization about climate mitigation policies. Commun Res 39(6):701–723. https://doi.org/10.1177/0093650211416646

Hart P, Feldman L, Leiserowitz A, Maibach E (2015a) Extending the impacts of hostile media perceptions: influences on discussion and opinion polarization in context of climate change. Sci Commun 37(4):506–532. https://doi.org/10.1177/1075547015592067

Hart P, Nisbet E, Myers T (2015b) Public attention to science and political news and support for climate change mitigation. Nat Clim Chang 5(6):541–545. https://doi.org/10.1038/nclimate2577

Heath Y, Gifford R (2006) Free-market ideology and environmental degradation: the case of belief in global climate change. Environ Behav 38(1):48–71. https://doi.org/10.1177/0013916505277998

Hmielowski J, Feldman L, Myers T, Leiserowitz A, Maibach E (2014) An attack on science? Media use, trust in scientists, and perceptions of global warming. Public Underst Sci 23(7):866–883. https://doi.org/10.1177/0963662513480091

Hoffman A (2015) How culture shapes the climate change debate. Stanford University Press, Redwood City

Holan A (2017) In context: what Rush Limbaugh said about hurricane Irma before evacuation. In: Punditfact. Available at: https://www.politifact.com/punditfact/article/2017/sep/10/context-what-rush-limbaugh-said-about-hurricane-ir/. Accessed 17 Apr 2018

Hornsey M, Harris E, Bain P, Fielding K (2016) Meta-analyses of the determinants and outcomes of belief in climate change. Nat Clim Chang 6(6):622–626. https://doi.org/10.1038/nclimate2943

Howe P, Mildenberger M, Marlon J, Leiserowitz A (2015) Geographic variation in opinions on climate change at state and local scales in the U S A. Nat Clim Chang 5(6):596–603. https://doi.org/10.1038/nclimate2583

Irfan U (2018) Trump White House issues climate change report undermining its own policy. In: Vox. Available at: https://www.vox.com/2018/11/26/18112505/national-climate-assessment-2018-trump. Accessed 5 Dec 2018

Jackson B (2011) Gingrich on climate change. In: FactCheck.org. Available at: https://www.factcheck.org/2011/12/gingrich-on-climate-change/. Accessed 2 Feb 2018

Jacques P, Knox C (2016) Hurricanes and hegemony: a qualitative analysis of micro-level climate change denial discourses. Environ Polit 25(5):831–852. https://doi.org/10.1080/09644016.2016.1189233

Jacques P, Dunlap R, Freeman M (2008) The organisation of denial: conservative think tanks and environmental scepticism. Environ Polit 17(3):349–385. https://doi.org/10.1080/09644010802055576

Joireman J, Truelove H, Duell B (2010) Effect of outdoor temperature, heat primes and anchoring on belief in global warming. J Environ Psychol 30(4):358–367. https://doi.org/10.1016/j.jenvp.2010.03.004

Jost J (2018) Underestimating belief in climate change. Nat Clim Chang 8(3):189–190. https://doi.org/10.1038/s41558-018-0082-4

Kahan D (2013) Making climate-science communication evidence-based—all the way down. In: Boykoff M, Crow D (eds) Culture, politics and climate change: how information shapes our common future. Routledge, London, pp 203–220. https://doi.org/10.2139/ssrn.2216469

Kahan D (2015) Climate-science communication and the measurement problem. Polit Psychol 36(S1):1–43. https://doi.org/10.1111/pops.12244

Kahan D (2016) The 'gateway belief' illusion: reanalyzing the results of a scientific-consensus messaging study. J Sci Commun 16(5):1–20. https://doi.org/10.2139/ssrn.2779661

Kahan D, Jenkins-Smith H, Braman D (2011) Cultural cognition of scientific consensus. J Risk Res 14(2):147–174. https://doi.org/10.1080/13669877.2010.511246

Kahan D, Peters E, Wittlin M, Slovic P, Ouellette L, Braman D, Mandel G (2012) The polarizing impact of scientific literacy and numeracy on perceived climate change risks. Nat Clim Chang 2(10):732–735

Kahan D, Jenkins-Smith H, Tarantola T, Silva C, Braman D (2015) Geoengineering and climate change polarization: testing a two-channel model of science communication. Ann Am Acad Pol Soc Sci 658(1):192–222. https://doi.org/10.1177/0002716214559002

Kellstedt P, Zahran S, Vedlitz A (2008) Personal efficacy, the information environment, and attitudes toward global warming and climate change in the United States. Risk Anal Int J 28(1):113–126. https://doi.org/10.1111/j.1539-6924.2008.01010.x

Konisky D, Hughes L, Kaylor C (2016) Extreme weather events and climate change concern. Clim Chang 134(4):533–547. https://doi.org/10.1007/s10584-015-1555-3

Kousser T, Tranter B (2018) The influence of political leaders on climate change attitudes. Glob Environ Chang 50:100–109. https://doi.org/10.1016/j.gloenvcha.2018.03.005

Kraft P, Lodge M, Taber C (2015) Why people "don't trust the evidence:" motivated reasoning and scientific beliefs. Ann Am Acad Pol Soc Sci 658(1):121–133. https://doi.org/10.1177/0002716214554758

Krosnick J, MacInnis B (2012) Trends in American public opinion on global warming policies between 2010 and 2012. Unpublished manuscript. Available at: https://climateaccess.org/system/files/Krosnick_Trends%20in%20American%20Public%20Opinion.pdf. Accessed 9 May 2018

Krosnick J, Holbrook A, Visser P (2000) The impact of the fall 1997 debate about global warming on American public opinion. Public Underst Sci 9(3):239–260. https://doi.org/10.1088/0963-6625/9/3/303

Krosnick J, Holbrook A, Lowe L, Visser P (2006) The origins and consequences of democratic citizens' policy agendas: a study of popular concern about global warming. Clim Chang 77(1–2):7–43. https://doi.org/10.1007/s10584-006-9068-8

Kunda Z (1990) The case for motivated reasoning. Psychol Bull 108(3):480–498. https://doi.org/10.1037/0033-2909.108.3.480

Landrum A, Baxter SE Jr, Shafto P (2015) Learning to trust and trusting to learn: a theoretical framework. Trends Cogn Sci 19(3):109–111. https://doi.org/10.1016/j.tics.2014.12.007

Landrum A, Lull R, Akin H, Hasell A, Jamieson K (2017) Processing the papal encyclical through perceptual filters: Pope Francis, identity-protective cognition, and climate change concern. Cognition 166:1–12. https://doi.org/10.1016/j.cognition.2017.05.015

Lee T, Markowitz E, Howe P, Ko C, Leiserowitz A (2015) Predictors of public climate change awareness and risk perception around the world. Nat Clim Chang 5(11):1014–1020. https://doi.org/10.1038/nclimate2728

Lehmann E (2016) Many more Republicans now believe in climate change: poll shows a big leap from two years ago. Scientific American Available at: https://www.scientificamericancom/article/many-more-republicans-now-believe-in-climate-change/. Accessed 3 June 2018

Leiserowitz A (2005) American risk perceptions: is climate change dangerous? Risk Anal Int J 25(6):1433–1442. https://doi.org/10.1111/j.1540-6261.2005.00690.x

Leiserowitz A (2006) Climate change risk perception and policy preferences: the role of affect, imagery, and values. Clim Chang 77(1–2):45–72. https://doi.org/10.1007/s10584-006-9059-9

Leiserowitz A (2007) Communicating the risks of global warming: American risk perceptions, affective images and interpretive communities. In: Moser S, Dilling L (eds) Creating a climate for change: communicating climate change –facilitating social change. Cambridge University Press, Cambridge, pp 44–63. https://doi.org/10.1017/CBO9780511535871.005

Leiserowitz A, Maibach E, Roser-Renouf C, Rosenthal S, Cutler M, Kotcher J (2018) Climate change in the American mind: March 2018. Yale Program on Climate Change Communication. Yale University and George Mason University, New Haven

Levendusky M (2013) Partisan media exposure and attitudes toward the opposition. Polit Commun 30(4):565–581. https://doi.org/10.1080/10584609.2012.737435

Levine A, Kline R (2017) A new approach for evaluating climate change communication. Clim Chang 142(1–2):301–309. https://doi.org/10.1007/s10584-017-1952-x

Lewandowsky S, Gignac G, Vaughan S (2013) The pivotal role of perceived scientific consensus in acceptance of science. Nat Clim Chang 3(4):399–404. https://doi.org/10.1038/nclimate1720

Li N, Hilgard J, Scheufele D, Winneg K, Jamieson K (2016) Cross-pressuring conservative Catholics? Effects of Pope Francis' encyclical on the US public opinion on climate change. Clim Chang 139(3–4):367–380. https://doi.org/10.1007/s10584-016-1821-z

Lorenzoni I, Pidgeon N (2006) Public views on climate change: European and USA perspectives. Clim Chang 77(1–2):73–95. https://doi.org/10.1007/s10584-006-9072-z

Lupia A (2013) Communicating science in politicized environments. Proc Nat Acad Sci U S A 110 (Supplement 3):14048–14054. Fischhoff B (ed). https://doi.org/10.1073/pnas.1212726110

Lupia A, McCubbins M (1998) The democratic dilemma: can citizens learn what they need to know? Cambridge University Press, Cambridge

Ma Y, Dixon G, Hmielowski J (2019) Psychological reactance from reading basic facts on climate change: the role of prior views and political identification. Environ Commun 13(1):71–86. https://doi.org/10.1080/17524032.2018.1548369

Maibach E, Leiserowitz A, Roser-Renouf C, Myers T, Rosenthal S, Feinberg G (2015) The Francis effect: how Pope Francis changed the conversation about global warming. Yale Program on Climate Change Communication. Yale University and George Mason University, New Haven.

Malka A, Krosnick J, Langer G (2009) The association of knowledge with concern about global warming: trusted information sources shape public thinking. Risk Anal 29(5):633–647. https://doi.org/10.1111/j.1539-6924.2009.01220.x

McCright A, Dunlap R (2000) Challenging global warming as a social problem: an analysis of the conservative movement's counter-claims. Soc Probl 47(4):499–522

McCright A, Dunlap R (2003) Defeating Kyoto: the conservative movement's impact on U.S. climate change policy. Soc Probl 50(3):348–373. https://doi.org/10.1525/sp.2003.50.3.348

McCright A, Dunlap R (2011a) The politicization of climate change and polarization in the American public's views of global warming, 2001–2010. Sociol Q 52(2):155–194. https://doi.org/10.1111/j.1533-8525.2011.01198.x

McCright A, Dunlap R (2011b) Cool dudes: the denial of climate change among conservative white males in the United States. Glob Environ Chang 21(4):1163–1172. https://doi.org/10.1016/j.gloenvcha.2011.06.003

McCright A, Charters M, Dentzman K, Dietz T (2016) Examining the effectiveness of climate change frames in the face of a climate change denial counter-frame. Top Cogn Sci 8(1):76–97. https://doi.org/10.1111/tops.12171

Meirick P, Nisbett G (2011) I approve this message: effects of sponsorship, ad tone, and reactance in 2008 presidential advertising. Mass Commun Soc 14(5):666–689. https://doi.org/10.1080/15205436.2010.530381

Mildenberger M, Tingley D (2017) Beliefs about climate beliefs: the importance of second-order opinions for climate politics. Br J Polit Sci:1–29. https://doi.org/10.1017/S0007123417000321

Mildenberger M, Lubell M, Hummel M (2019) Personalized risk messaging can reduce climate concerns. Glob Environ Chang 55:15–24. https://doi.org/10.1016/j.gloenvcha.2019.01.002

Milfont T (2012) The interplay between knowledge, perceived efficacy, and concern about global warming and climate change: a one-year longitudinal study. Risk Anal Int J 32(6):1003–1020. https://doi.org/10.1111/j.1539-6924.2012.01800.x

Moser S (2010) Communicating climate change: history, challenges, process and future directions. Wiley Interdiscip Rev Clim Chang 1(1):31–53. https://doi.org/10.1002/wcc.11

Moser S, Dilling L (2011) Communicating climate change: closing the science-action gap. In: Dryzek J, Norgaard R, Schlosberg (eds) The Oxford handbook of climate change and society, pp 161–174. https://doi.org/10.1093/oxfordhb/9780199566600.003.0011
Motel S (2014) Polls show most Americans believe in climate change, but give it low priority. In: Pew Research Center. Available at: https://www.pewresearch.org/fact-tank/2014/09/23/most-americans-believe-in-climate-change-but-give-it-low-priority/. Accessed 20 Nov 2018
Motta M (2018) The polarizing effect of the march for science on attitudes toward scientists. Polit Sci Polit 51(4):782–788. https://doi.org/10.1017/S1049096518000938
Myers T, Nisbet M, Maibach E, Leiserowitz A (2012) A public health frame arouses hopeful emotions about climate change. Clim Chang 113(3–4):1105–1112. https://doi.org/10.1007/s10584-012-0513-6
Nazarvan A (2017) Rush Limbaugh on hurricane Irma: powerful storm or liberal conspiracy? In: Newsweek. Available at: https://www.newsweek.com/irma-hurricane-limbaugh-climate-change-liberal-conspiracy-trump-deep-state-659890. Accessed 8 June 2018
Nerlich B, Koteyko N, Brown B (2010) Theory and language of climate change communication. Wiley Interdiscip Rev Clim Chang 1(1):97–110. https://doi.org/10.1002/wcc.2
Nisbet M (2009) Communicating climate change: why frames matter for public engagement. Environ Sci Policy Sustain Dev 51(2):12–23. https://doi.org/10.3200/ENVT.51.2.12-23
Nisbet M (2011) Climate shift: clear vision for the next decade of public debate. https://grist.files.wordpress.com/2011/04/climateshift_report_americanuniversity-updated.pdf. Accessed 12 Apr 2018
Nisbet M (2014) Engaging in science policy controversies: insights from the U.S. debate over climate change. In: Bucchi M, Trench B (eds) Handbook of the public communication of science and technology, 2nd edn. Routledge, London, pp 173–185
Nisbet M, Myers T (2007) The polls—trends: twenty years of public opinion about global warming. Public Opin Q 71(3):444–470. https://doi.org/10.1093/poq/nfm031
Nisbet E, Cooper K, Garrett R (2015) The partisan brain: how dissonant science messages lead conservatives and liberals to (dis) trust science. Ann Am Acad Pol Soc Sci 658(1):36–66. https://doi.org/10.1177/0002716214555474
Nyhan B, Reifler J (2010) When corrections fail: the persistence of political misperceptions. Polit Behav 32(2):303–330. https://doi.org/10.1007/s11109-010-9112-2
O'Connor R, Bard R, Fisher A (1999) Risk perceptions, general environmental beliefs, and willingness to address climate change. Risk Anal 19(3):461–471. https://doi.org/10.1111/j.1539-6924.1999.tb00421.x
O'Keefe D (2012) Persuasion. In: Donsbach W (ed) The international encyclopedia of communication. https://doi.org/10.1002/9781405186407.wbiecp031.pub2
O'Neill S (2013) Image matters: climate change imagery in US, UK and Australian newspapers. Geoforum 49:10–19. https://doi.org/10.1016/j.geoforum.2013.04.030
O'Neill S, Nicholson-Cole S (2009) Fear won't do it: promoting positive engagement with climate change through visual and iconic representations. Sci Commun 30(3):355–379. https://doi.org/10.1177/1075547008329201
Oreskes N, Conway E (2010) Merchants of doubt: how a handful of scientists obscured the truth on issues from tobacco smoke to global warming. Bloomsbury Publishing, London
Palm R, Lewis G, Feng B (2017) What causes people to change their opinion about climate change? Ann Am Assoc Geogr 107(4):883–896. https://doi.org/10.1080/24694452.2016.1270193
Palmer E (2018) Rush Limbaugh says hurricane Florence forecast is trying to 'heighten the belief in climate change'. In: Newsweek. Available at: https://www.newsweek.com/rush-limbaugh-says-hurricane-florence-forecast-trying-heighten-belief-climate-1117416. Accessed 14 Sept 2018
Peterson C (1989) Experts, OMB spar on global warming: "greenhouse effect" may be accelerating, scientists tell hearing. In: Washington Post. Available at: https://www.washingtonpost.com/archive/politics/1989/05/09/experts-omb-spar-on-global-warming/7fc219d4-4693-4663-bf83-319fa97d7dda/?utm_term=.821af1e8990e. Accessed 28 June 2018

Pew Research Center (2018) Majorities see government efforts to protect the environment as insufficient. Available at: http://www.pewresearch.org/science/2018/05/14/majorities-see-government-efforts-to-protect-the-environment-as-insufficient/

Priest S (2016) Communicating climate change: the path forward. Palgrave Macmillan, Basingstoke

Ranney M, Clark D (2016) Climate change conceptual change: scientific information can transform attitudes. Top Cogn Sci 8(1):49–75. https://doi.org/10.1111/tops.12187

Reser J, Bradley G, Ellul M (2014) Encountering climate change: 'seeing' is more than 'believing'. Wiley Interdiscip Rev Clim Chang 5(4):521–537. https://doi.org/10.1002/wcc.286

Retchless D (2018) Understanding local sea-level rise risk perceptions and the power of maps to change them: the effects of distance and doubt. Environ Behav 50(5):483–511. https://doi.org/10.1177/0013916517709043

Rudman L, McLean M, Bunzl M (2013) When truth is personally inconvenient, attitudes change: the impact of extreme weather on implicit support for green politicians and explicit climate-change beliefs. Psychol Sci 24(11):2290–2296. https://doi.org/10.1177/0956797613492775

Saad L, Jones J (2016) US concern about global warming at eight-year high. In: Gallup Politics. Available at: https://news.gallup.com/poll/190010/concern-global-warming-eight-year-high.aspx. Accessed 4 Apr 2018

Scannell L, Gifford R (2013) Personally relevant climate change: the role of place attachment and local versus global message framing in engagement. Environ Behav 45(1):60–85. https://doi.org/10.1177/0013916511421196

Schuldt J, Pearson A (2016) The role of race and ethnicity in climate change polarization: evidence from a US national survey experiment. Clim Chang 136(3–4):495–505. https://doi.org/10.1007/s10584-016-1631-3

Schuldt J, Roh S (2014) Of accessibility and applicability: how heat-related cues affect belief in "global warming" versus "climate change". Soc Cogn 32(3):217–238. https://doi.org/10.1521/soco.2014.32.3.217

Schuldt J, Konrath S, Schwarz N (2011) "Global warming" or "climate change"? Whether the planet is warming depends on question wording. Public Opin Q 75(1):115–124. https://doi.org/10.1093/poq/nfq073

Schuldt J, Roh S, Schwarz N (2015) Questionnaire design effects in climate change surveys: implications for the partisan divide. Ann Am Acad Pol Soc Sci 658(1):67–85. https://doi.org/10.1177/0002716214555066

Schuldt J, Enns P, Cavaliere V (2017) Does the label really matter? Evidence that the US public continues to doubt "global warming" more than "climate change". Clim Chang 143(1–2):271–280. https://doi.org/10.1007/s10584-017-1993-1

Schweizer S, Davis S, Thompson J (2013) Changing the conversation about climate change: a theoretical framework for place-based climate change engagement. Environ Commun J Nat Cult 7(1):42–62. https://doi.org/10.1080/17524032.2012.753634

Severson A, Coleman E (2015) Moral frames and climate change policy attitudes. Soc Sci Q 96(5):1277–1290. https://doi.org/10.1111/ssqu.12159

Shao W, Keim B, Garand J, Hamilton L (2014) Weather, climate, and the economy: explaining risk perceptions of global warming, 2001–10. Weather Clim Soc 6(1):119–134. https://doi.org/10.1175/WCAS-D-13-00029.1

Shear M (2017) Trump will withdraw US from Paris climate agreement. In: The New York Times. Available at: https://nyti.ms/2rv52tR. Accessed 26 Jan 2018

Slimak M, Dietz T (2006) Personal values, beliefs, and ecological risk perception. Risk Anal 26(6):1689–1705. https://doi.org/10.1111/j.1539-6924.2006.00832.x

Slovic P (1999) Trust, emotion, sex, politics, and science: surveying the risk-assessment battlefield. Risk Anal Int J 19(4):689–701. https://doi.org/10.1111/j.1539-6924.1999.tb00439.x

Smith N, Leiserowitz A (2012) The rise of global warming skepticism: exploring affective image associations in the United States over time. Risk Anal Int J 32(6):1021–1032. https://doi.org/10.1111/j.1539-6924.2012.01801.x

Spence A, Pidgeon N (2010) Framing and communicating climate change: the effects of distance and outcome frame manipulations. Glob Environ Chang 20(4):656–667. https://doi.org/10.1016/j.gloenvcha.2010.07.002

Sundblad E, Biel A, Gärling T (2007) Cognitive and affective risk judgements related to climate change. J Environ Psychol 27(2):97–106. https://doi.org/10.1016/j.jenvp.2007.01.003

Sussman G, Byron D (2013) US politics and climate change: science confronts policy. Lynne Rienner Publishers, Boulder

Taber C, Lodge M (2006) Motivated skepticism in the evaluation of political beliefs. Am J Polit Sci 50(3):755–769. https://doi.org/10.1111/j.1540-5907.2006.00214.x

Tesler M (2018) Elite domination of public doubts about climate change (not evolution). Polit Commun 35(2):306–326. https://doi.org/10.1080/10584609.2017.1380092

Ungar S (1992) The rise and (relative) decline of global warming as a social problem. Sociol Q 33(4):483–501. https://doi.org/10.1111/j.1533-8525.1992.tb00139.x

Van Boven L, Ehret P, Sherman D (2018) Psychological barriers to bipartisan public support for climate policy. Perspect Psychol Sci 13(4):492–507. https://doi.org/10.1177/1745691617748966

van der Linden S (2015) The social-psychological determinants of climate change risk perceptions: towards a comprehensive model. J Environ Psychol 41:112–124. https://doi.org/10.1016/j.jenvp.2014.11.012

van der Linden S (2016) A conceptual critique of the cultural cognition thesis. Sci Commun 38(1):128–138. https://doi.org/10.1177/1075547015614970

van der Linden S (2017) Determinants and measurement of climate change risk perception, worry, and concern. Oxford research encyclopedia of climate science. https://doi.org/10.1093/acrefore/9780190228620.013.318

van der Linden S, Leiserowitz A, Feinberg G, Maibach E (2015a) The scientific consensus on climate change as a gateway belief: experimental evidence. PLoS One 10(2):e0118489. https://doi.org/10.1371/journal.pone.0118489

van der Linden S, Maibach E, Leiserowitz A (2015b) Improving public engagement with climate change: five "best practice" insights from psychological science. Perspect Psychol Sci 10(6):758–763. https://doi.org/10.1177/1745691615598516

van der Linden S, Leiserowitz A, Maibach E (2017) Gateway illusion or cultural cognition confusion? J Sci Commun 16(05). https://doi.org/10.22323/2.16050204

van der Linden S, Leiserowitz A, Maibach E (2019) The gateway belief model: a large-scale replication. J Environ Psychol 62:49–58. https://doi.org/10.1016/j.jenvp.2019.01.009

Villar A, Krosnick J (2011) Global warming vs. climate change, taxes vs. prices: does word choice matter? Clim Chang 105(1–2):1–12. https://doi.org/10.1007/s10584-010-9882-x

Walsh L (2015) The visual rhetoric of climate change. Wiley Interdiscip Rev Clim Chang 6(4):361–368. https://doi.org/10.1002/wcc.342

Weber E (2010) What shapes perceptions of climate change? Wiley Interdiscip Rev Clim Chang 1(3):332–342. https://doi.org/10.1002/wcc.41

Weber E, Stern P (2011) Public understanding of climate change in the United States. Am Psychol 66(4):315–328. https://doi.org/10.1037/a0023253

Weeks B, Garrett R (2014) Electoral consequences of political rumors: motivated reasoning, candidate rumors, and vote choice during the 2008 US presidential election. Int J Publ Opin Res 26(4):401–422. https://doi.org/10.1093/ijpor/edu005

West D, Orr M (2007) Race, gender, and communications in natural disasters. Policy Stud J 35(4):569–586. https://doi.org/10.1111/j.1541-0072.2007.00237.x

Whitmarsh L (2008) Are flood victims more concerned about climate change than other people? The role of direct experience in risk perception and behavioural response. J Risk Res 11(3):351–374. https://doi.org/10.1080/13669870701552235

Wolsko C, Ariceaga H, Seiden J (2016) Red, white, and blue enough to be green: effects of moral framing on climate change attitudes and conservation behaviors. J Exp Soc Psychol 65:7–19. https://doi.org/10.1016/j.jesp.2016.02.005

Zaval L, Keenan E, Johnson E, Weber E (2014) How warm days increase belief in global warming. Nat Clim Chang 4(2):143. https://doi.org/10.1038/nclimate2093

Zhou J (2016) Boomerangs versus javelins: how polarization constrains communication on climate change. Environ Polit 25(5):788–811. https://doi.org/10.1080/09644016.2016.1166602

Zia A, Todd A (2010) Evaluating the effects of ideology on public understanding of climate change science: how to improve communication across ideological divides? Public Underst Sci 19(6):743–761. https://doi.org/10.1177/0963662509357871

Chapter 4
The South Florida Study Area

Abstract South Florida is a popular destination for new residents and tourists, attracting a great deal of residential property investment from other parts of the United States as well as Canada and Latin America. Its flat coastal topography, rarely rising above 10 meters above sea-level, makes it susceptible to both storm surge and tidal flooding. The limestone bedrock underlying many of the urban areas adds to the flood risks. Settlement of Florida followed rail and interstate connection, the draining of former wetlands to create residential property, and the development of air conditioning to facilitate year-round occupation. The specific portions of Florida studied in the research reported here include postal code zones (zip-codes) in the urban areas of Tampa-Saint Petersburg-Clearwater, Fort Myers, Key West, Miami-Dade County, Fort Lauderdale, West Palm Beach and Palm Beach, and Vero Beach. Most of these areas experienced real estate speculation, with accompanying boom-and-bust periods, as well as numerous home sales to non-resident owners who do not have the same connection to the region as owner-occupiers. Despite repeated flooding, Floridians have shown an ambivalence about their susceptibility to sea-level rise.

Keywords Settlement of South Florida · Real estate speculation · Finger islands · Urban flooding

South Florida has been transformed by a combination of air conditioning and water management into a "subtropical paradise" for residents and tourists (Grunwald 2017). Buyers from all income groups flock to South Florida, from the top 1% of the income distribution to the bottom 20%, from the United States and abroad. Some of the wealthy U.S. buyers are said to be motivated by low taxes in Florida, including the lack of a state income tax, compared with states like New York, New Jersey and California (De Lea 2019). In addition, about one-fifth of the home purchases totaling almost $23 billion in value are made by foreign buyers, primarily from Canada and Latin America. Of these, two-thirds purchase the property for vacation or residential rental rather than owner-occupancy (Florida Realtors/National Association of Realtors 2018), and the majority of these purchases are in the Miami-Fort Lauderdale-West Palm Beach area.

R. Palm, T. Bolsen, *Climate Change and Sea Level Rise in South Florida*, Coastal Research Library 34, https://doi.org/10.1007/978-3-030-32602-9_4

4.1 The Physical Setting: Topography and Water in South Florida

Florida's coast has changed dramatically as a function of rising and falling sea-levels, which in turn are related to glaciation. During the last interglacial era about 120,000 years ago, when temperatures were warmer and the sea-levels were higher, the land area of Florida was much smaller than it is at present. By contrast, the early human settlement in the state was established about 12,000 years on a Florida peninsula that was larger, cooler and dryer than at-present. The current coastal boundary emerged about 3000 B.C.E. (Bostick et al. 2018). Most of urbanized South Florida is flat, with very little land rising above 10 meters (32.8 feet) above sea-level (Fig. 4.1). For this reason, most areas are susceptible to a relatively small variation in sea-level, especially when sea-level rise is combined with storm surge and tidal flooding.

Another characteristic of the Florida coast is the nature of the limestone bedrock on which the urban areas have been superimposed. During most of its history, southern Florida was submerged under the ocean, resulting in the accumulation of a thick layer of limestone. This provides the amenity of coral reefs and marine vegetation for divers to explore, but also creates a porous limestone base for construction. Porous limestone is particularly susceptible to sudden flooding because the limestone, as opposed to sand or silt, does not absorb and diffuse rain or storm surge

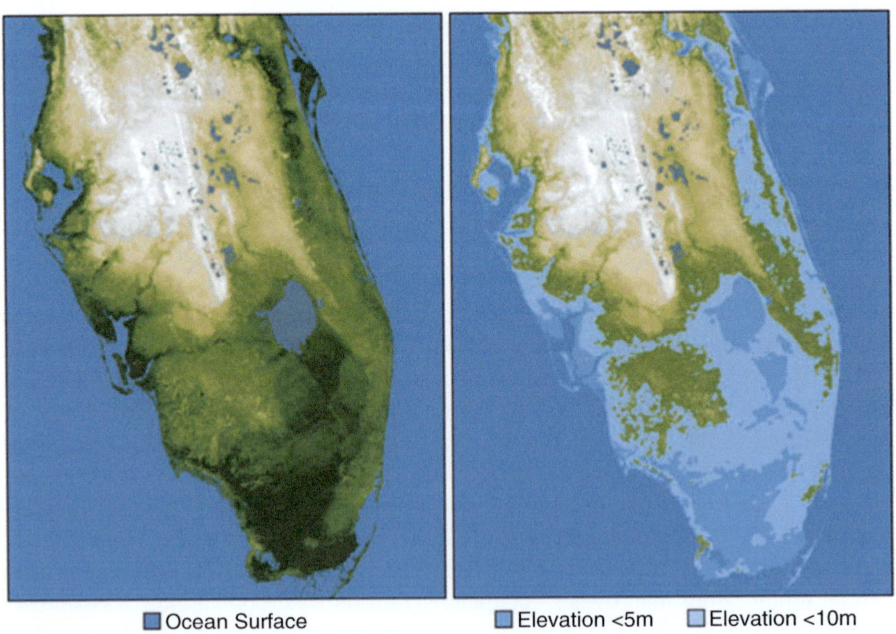

Fig. 4.1 Current coastline and coastline with sea-level rise. (Image Courtesy SRTM Team NASA/JPL/NIMA)

water. Instead, because the limestone can absorb so little additional water, groundwater levels can rise during periods of intense rain downpours or storm surge. Floodwaters are forced back to the surface and the groundwater levels can rise more dramatically than would be expected along rivers or coastlines (Naughton et al. 2018). The combination of the limestone base and low elevation make this region particularly susceptible to flooding, especially during periods of intense rainfall. According to data from the National Weather Service (Fisk 2019), annual precipitation in Miami in the last hundred years varied from a low of 81.28 centimeters (32 inches) to a high of almost 2.286 meters (90 inches), with an average of about 1.524 meters (60 inches) per year. Tropical cyclones have dropped as much as 35.56 centimeters (14 inches) in a 24-h period. The result is susceptibility to urban flooding even at current sea levels.

Florida's cities have thus been built on a fragile environmental base, subject to flooding and at the same time, perhaps ironically, to shortages of fresh water as saltwater intrudes into urban water sources. How did such an area come to be settled?

4.2 Settlement of South Florida

Although human settlement in Florida dates from early Native American inhabitation some 12,000 years ago (Purdy 2008), European contact began with Spanish exploration in the early 1500s (Milanich 1998). The development of an urbanized South Florida began in earnest after Spain ceded the land to the United States in 1821, Florida became an Organized Incorporated Territory in 1822, and became a state in 1845.

South Florida generally has been marked by the purchase of tracts of land by entrepreneurs from other parts of the country, the extension of transportation lines – notably, rail service linking the state to the rest of the national rail grid, the promotion of tourism, and the development of settlements with a large associated real estate industry. The first of a series of real estate bubbles in Florida occurred in the 1920s. Drainage of the Everglades wetlands combined with the connection of the area to the national railroad grid encouraged the development of many projects, fueled by easy credit for buyers, speculators, and the promise of appreciating property values. A combination of disastrous hurricanes and the general economic depression of the late 1920s halted the speculation in real estate, at least for a few years.

One of the building practices that makes parts of coast Florida particularly vulnerable to tidal flooding was the dredge-and-fill creation of artificial islands or peninsulas along canals created to give each part of the subdivision direct access to water. Draining coastal wetlands created thousands of miles of artificial canals (Nicholls and Crompton 2017). This new land was particularly susceptible to tidal and storm surge flooding, but at the same time provided huge profits for speculators by creating housing tracts with a water view and also direct access from the home by boat to coastal water (Fig. 4.2).

Fig. 4.2 Creation of Land Through Dredge-and-Fill. (Source: U.S. National Archives and Records Administration)

South Florida continues to recover from the most recent housing bubble of 2007–2009 that affected many U.S. cities (Baker 2018). This bubble occurred as a result of multiple factors, including relaxed lending standards and unscrupulous practices by investors, lenders, and mortgage brokers and those who secured these mortgages. When the bubble burst, the Florida housing market was hit hard, with many homeowners figuratively "under water," in the sense that their mortgages were worth more than the sales prices of their homes.

South Florida remains a vulnerable "paradise," promoted by those seeking to profit through tourism and real estate sales, but also susceptible to environmental hazards associated with the very characteristics that make it attractive: its topography, its access to the sea, its water resources, and its semi-tropical climate that brings both glorious winter sunshine and devastating summer and autumn hurricanes.

4.3 Population and Economic Characteristics of the Study Sites

There is no single accepted boundary for South Florida, but for this study, we defined the region as south of the 28th parallel north. Respondents were surveyed in the Tampa-Saint Petersburg-Clearwater area, Fort Myers, Key West, Miami-Dade County, Fort Lauderdale, West Palm Beach and Palm Beach, and Vero Beach. Since the purpose of the empirical study was to ascertain the attitudes of coastal residents of South Florida, we restricted our selection of study areas to residents of 166 postal (zip) codes in these urbanized areas (Fig. 4.3).

These sites vary greatly in population size, from the very small Vero Beach and Key West, to the large metropolitan complexes in the Tampa Bay region and Naples/ Fort Myers on the Gulf Coast of Florida and the Miami-Dade, Fort Lauderdale and West Palm Beach areas on the Atlantic coast (Table 4.1). The municipalities vary widely in age, ethnic and racial composition, and economic diversity.

The study areas have much in common topographically but range widely with respect to their current demographic make-up. This variation is in part linked to the ways in which the areas were settled.

Key West, with a population of just over 25,000, is primarily a tourist destination. Like many other parts of South Florida, it grew slowly during the nineteenth century until it was connected to the rest of the country by rail, in this case Henry Flagler's Florida East Coast Railway. A series of land booms and increased tourist trade fueled continued development, with the first highways completed in 1938, connecting Key West to the rest of Florida via U.S. Route 1 (Ogle 2003). The economy continued to grow through a combination of the military importance of Key West's location and tourism. Famous residents included Ernest Hemingway, Robert Frost, Judy Blume and Tennessee Williams. The influence of the arts continues as well as its reputation as inclusive: "its embrace of diversity and willingness to set itself apart from the Florida mainland goes back to its origins as a haven for an eclectic blend of Native Americans, Spanish, freed slaves, seafarers (including

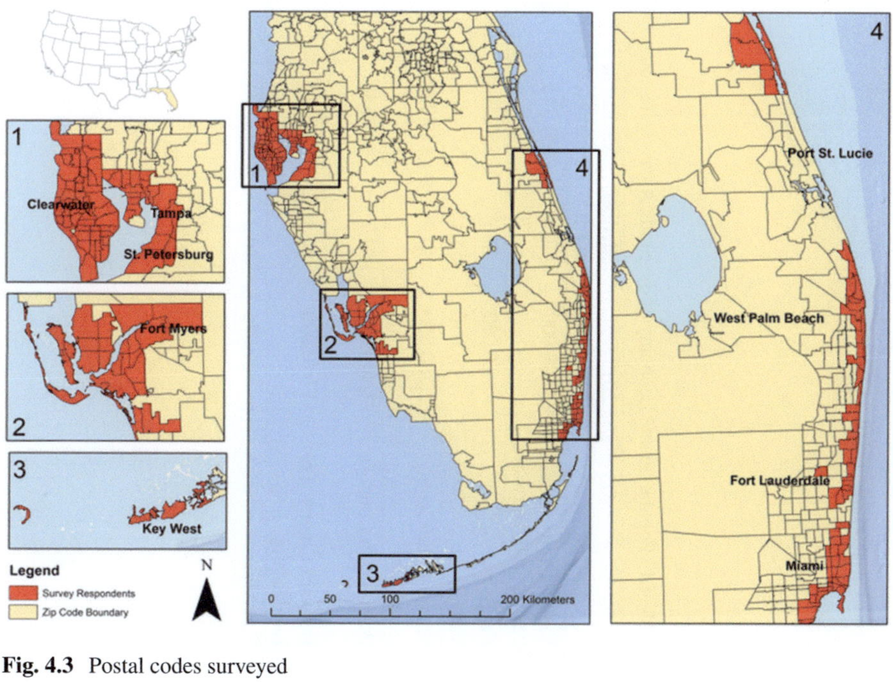

Fig. 4.3 Postal codes surveyed

Table 4.1 Population and Economy of the Study Areas (Data compiled by DataUSA.ib)

	2017 Population	Median age	Median household income, 2017	% Poverty	Primary industries	% White	%Hispanic	% Black
Tampa-Saint Petersburg-Clearwater	3,003,000	42.3	$51,115	14.2%	Healthcare, retail, accommodation & food service	63.8%	18.6%	11.3%
Key West	26,039	40.6	$58,494	12.7%	Accommodation & food service	64.1%	21.6%	11.3%
Vero Beach	16,070	54.1	$38,405	19.9%	Retail, healthcare	83.9%	8.7%	11.3%
Miami-Dade County	2,700,000	39.9	$45,935	18.3%	Healthcare, retail, accommodation & food service	13.6%	67.7%	16.1%
Palm Beach County	1,440,000	44.7	$57,580	12.5%	Healthcare, retail, accommodation & food service	55.8%	21.5%	18.2%
Broward County (Fort Lauderdale)	1,910,000	40.3	$54,212	13.5%	Healthcare, retail, accommodation & food service	37.4%	28.7%	27.6%
Fort Myers	71,051	40.2	$38,971	22.6%	Healthcare, retail	47.8%	21.4%	27.2%

pirates), queers, and creatives" (Artavia 2018). Politically, Key West stands out within Monroe county as a liberal and Democratic-oriented community: in the 2016 presidential election, President Trump won Monroe county, but Secretary Clinton won Key West. In addition, the historically black neighborhood within Key West (Bahama Village) went overwhelmingly for Clinton. Given this background, we would expect coastal residents to be more likely to be conscious of climate change and their vulnerability to sea-level rise.

Tampa-Saint Petersburg-Clearwater, with a population of more than 4 million, is a conglomerate of several smaller areas surrounding the Tampa Bay. The census Metropolitan Statistical Area (MSA) includes Hillsborough and Pinellas Counties, as well as Hernando and Pasco Counties to the north. The City of Tampa began to grow in the 1880s when Henry Bradley Plant, a Connecticut investor, built a railroad linking Tampa with Charleston and Savannah, and established steamship service to both Cuba and Key West (Sammons 2010). The city grew from a population of 800 in 1880 to over 15,000 in 1900, its economy based on the mining of phosphates and the manufacture of cigars. These industries attracted immigrants from Cuba, Spain, and Italy. Like other cities in Florida, Tampa experienced land booms and busts, and had its share of interesting characters such as crime boss Charlie Wall and mafioso Santo Trafficante (Kerstein 2001). Construction of what later became the Tampa International Airport began during World War II when Tampa became a center for military operations. Rapid population growth took place in the 1950s and 1960s with the development of the tourist industry including the Busch Gardens attraction, and the establishment of the University of South Florida, now with over 40,000 students. Downtown Tampa remains a hub of a great deal of development. It is a major shipping port and is also second only to Miami as a cruise ship port. Tampa's population of more than 385,000 is diverse with more than 25% of the total population Black or African-American and also more than 25% Hispanic or Latino. In 2016, both the city of Tampa and Hillsborough County voted overwhelmingly for the Democratic presidential candidate, Hillary Clinton.

Across the Tampa Bay is the city of Saint Petersburg founded in 1875 by other Yankee investors, John C. Williams of Detroit and Peter Demens, a Russian immigrant who brought the Orange Belt Railway to the town. By the turn of the twentieth century, a wholesale fish industry had been established, and by the 1920s a large tourist trade fueled the economy. The population swelled during World War II because St. Petersburg was used as a training base for both the army and the army air force. The population continued to expand in the 1950s with the construction of the Sunshine Skyway Bridge in 1954 linking St. Petersburg with Bradenton, southward across the Tampa Bay. In the 1970s, it became a home base for major league baseball teams' winter training camps and started to become a retirement destination (Arsenault 1996). The central business district hosts corporate headquarters for Raymond James, the Home Shopping Network and Publix Supermarkets among others, as well as many galleries and museums including the Salvador Dali museum and the Florida Holocaust museum. Crime rates in St. Petersburg are also high, with the city ranking in the lowest 5th percentile among cities in the United States (NeighborhoodScout.com 2019). The population of St. Petersburg is over

263,000 with about 70% of the population being classified as white. Although Hillary Clinton won handily in the city of St. Petersburg itself, she lost in Pinellas county to Donald Trump in 2016.

The third city in this metropolitan area is Clearwater, with a population of just over 115,000 in 2017. Clearwater, Clearwater Beach (across the intercoastal waterway), Palm Harbor, Treasure Island and Largo are known for their beaches. Like other settlements in this region, Clearwater grew as a result of the extension of a railroad, this time by Henry B. Plant who built a resort hotel there in 1897. It became a tourist destination but was also a military training base during World War II. Clearwater is home of a marine aquarium as well as the headquarters of the Church of Scientology. Its population is firmly Republican, with Donald Trump winning most of the precincts in the 2016 presidential election.

Further south along the Gulf coast is the Fort Myers area including Cape Coral, the largest city in Lee County (population of over 180,000 in 2017). Fort Myers, with a current population of about 80,000, was originally established in 1821 when Florida became a U.S. territory as a fort on the Caloosahatchee River to protect settlers during the Seminole Indian Wars (Reaves 2007). The military installation itself was re-occupied by Union troops at the end of the Civil War and was later abandoned. Fort Myers was revived when Thomas Alva Edison built his home and his laboratory as a research site for goldenrod rubber. Edison became the city's most famous resident and convinced his friend Henry Ford to establish a winter residence there as well. In 1898 the Royal Palm hotel was opened, and the Atlantic Coast Line Railroad opened in 1904, making Fort Myers more accessible as a winter resort. The estates of Edison and Ford remain major tourist attractions. Lee County is firmly Republican, with Donald Trump picking up almost 60% of the county vote in the 2016 election.

Cape Coral began (Fig. 4.4) as a planned community, founded in 1957 by Leonard and Jack Rosen from Baltimore. They used what was at the time a very innovative marketing model, selling homesites by hosting dinner parties at local hotels and, eventually, flying buyers to inspect the properties as a means of selling land. Property was sold with the promise that this was "America's land of tomorrow, just $20 down and $20 a month for a quarter-acre slice of heaven" (Grunwald 2007, 2017). The developers aimed for an elevation of 1.676 meters (5.5 feet) for construction of housing, and therefore had to dredge the land to provide fill (Williams 2017a, b). The result was a series of 643.738 kilometers (400 miles) of canals with waterfront amenity property, but no initial plans for offices, shopping or even sewer and water service (Bernard and Hobe 1983; Dodrill 1983). Since Cape Coral originated as drained swampland, it remains susceptible to storm surge (Williams 2017c). Grunwald (2017) describes it this way: "Cape Coral is the ultimate microcosm of Florida. It's literally a peninsula jutting off the peninsula, the least natural, worst-planned, craziest-growing piece of an unnatural, badly planned, crazy-growing state."

The northernmost study site on the Atlantic coast is Vero Beach and Indian River County, places that were formed out of the existing Fort Pierce. Growth in this region was made possible by Henry Flagler's Florida's East Coast Railway, which first connected to Vero Beach in 1893. Land developers purchased large tracts of

Fig. 4.4 Cape Coral, Florida. (Source: Southwest Florida Regional Planning Council)

land to create cities, helped by the drainage of wetlands through the Land Reclamation Project (Meindl 2000). The economy had been based on citrus production as well as livestock ranching and timber. But the urbanized area was created by dredging canals, opening roads, and subdividing the land into building lots. Between 1914 and 1930, 643,738 kilometers (400 miles) of canals were dredged (Indian River County Administration 2019). The city and county were created in 1925 in order to escape from the jurisdiction of the "blue laws" of St. Lucie, forbidding commerce or the showing of movies on Sundays. After World War II, Vero Beach became the winter training ground of the Brooklyn Dodgers, and in the late 1950s became the home of Piper Aircraft (Miller 2017). Vero Beach expanded with the construction of bridges connecting the barrier island to the mainland and is now the site of numerous homes for sale at prices exceeding $5 million. Tourism has also recently been stimulated by the opening of Disney's Vero Beach Resort which boasts sea turtle nesting grounds as well as a Pirate's Plunge. The county is solidly Republican, with Donald Trump winning 61% of the vote in 2016.

The Miami region too was the product of the extension of the Florida East Coast Railway by Henry Flagler (Nijman 2011). The railroad eventually extended to Key West and transported agricultural products from Florida *to the north* as well as tourists *to the south*. Flagler purchased large tracts of land in Miami and developed its grid pattern, designed its basic infrastructure and created a separate area for black workers (Nijman 2011). Miami Beach, a project of John Collins from New Jersey and Carl Fisher of Indiana, was developed in the early part of the twentieth century, spurred by the construction of a bridge from Miami to Miami Beach (Foster

2000; Frazer and Guthrie Jr 1995). Settlement was made possible by draining the wetlands of the Florida Everglades, and as in other parts of South Florida, real estate was aggressively marketed.

The region is defined by the US Census as a Metropolitan Statistical Area including the counties of Miami-Dade, Broward, and Palm Beach, with a population in 2018 of 6.2 million residents. Indeed, the Census Bureau has included an even wider region based on commuting patterns that extends to Martin, St. Lucie, Indian River and Okeechobee counties with a total combined population of over 6.7 million. The major metropolitan areas are Miami-Miami Beach Kendall in Miami Dade County, Fort Lauderdale-Pompano Beach-Deerfield Beach in Broward County and West Palm Beach-Boca Raton-Boynton Beach in Palm Beach County. In contrast to the more conservative counties to the north and west, this region is reliably Democrat in orientation, particularly in Broward county.

The complexity of influences on development has created a large and low-density region made up of neighborhoods and municipalities that are diverse in ethnicity and income, in national origin and language, and are also highly segregated residentially. Nijman (2011) classified the residents of Miami neighborhoods as belonging to one of three categories: *locals* (the 20% who actually consider Miami as "home," and who may be unable to move because of poverty), *exiles* (those who are located in Miami but with the hope of returning to their homeland such as Cubans, Nicaraguans, Haitians, Venezuelans and others), and *mobiles* (mobile and affluent populations, sometimes international in origin, who may live in South Florida temporarily as on vacation or who establish one of several residences in the area). The spatial concentration of these groups has led to the development of a patchwork of communities, sometimes adjacent to one another, but with very different population characteristics. Nijman defines areas with high concentrations of "locals" as including Liberty City, Little Haiti, Overtown and Opa-locka, areas with high concentrations of "exiles" as Little Havana, Sweetwater and Hialeah, and areas of "mobiles" as including South Beach, Brickell, Fort Lauderdale and Key Biscayne (Nijman 2011).

Over the past 100 years, Miami has experienced periods of rapid development and then stagnation (Gridics staff 2016). The first boom was in the 1920s, fueled partly by the desire of wealthy people from the northern parts of the United States to have a winter home ("snowbirds"), partly by its new accessibility by rail, and partly because of the spread of the automobile, making family vacations by car feasible.

Successful developments often chose an exotic name or image and included "themed" street names. An example is Opa-Locka, the "Baghdad of Dade County", developed by Glenn Curtiss to "embody the look and spirit of illustrations he had seen in a copy of the *Arabian Nights*" (Derr 1989). Buildings were decorated with "minarets, domes, and assorted other psudo-Moorish accoutrements," and men were employed to dress as Saracens on Arabian hourses "to chase veiled women through the streets of his city." (Derr 1989). In a similar vein, George Merrick laid out Coral Gables to include "little villages" that "evoked places he'd never seen—China, the French countryside, the Netherlands, and he decreed that the houses there should match the distinctive styles of those nations." (Derr 1989; Parks 2015).

This real estate boom was not only based on attracting new residents from northern states, but also on luring investors wanting to make a quick profit: "for a small down payment, these real estate operators would buy a "binder" or 10% deposit notes on a lot and resell it for a quick profit, kiting prices skyward." (Cumming 2006). As this frenzy of speculation increased, the value of Florida real estate reached "unsustainable levels and were merely a pyramiding of paper profits" (Cumming 2006).

Part of the attraction of Miami in the early twentieth century was the combination of gambling and bootlegging of liquor, attracting such infamous residents as John Gotti, Meyer Lansky, and Al Capone. Smuggling, prostitution and other mob activities seemed to fit comfortability with other businesses in which the mob was engaged, including real estate and hotels (Nijman 2011).

World War II saw the creation of military districts to defend against German U-boats, and after the war there was a second housing boom that lasted through the 1950s. This housing boom was not only due to pent-up demand for housing by returning veterans, but also by the availability of air conditioning and the construction of new highways, making Miami more accessible and more attractive during the summer months. In addition, the migration of Cubans following the 1959 revolution added to the middle-class demand for housing.

Jewish settlement in South Florida had an early origin but began to take on steam in the 1920s and 1930s. Restrictive covenants on property made it difficult for Jews to settle in particular locations until a 1949 law ended real estate discrimination (Zerivitz 2019). Large influxes followed through the next decades, with many employed in the tourism or construction/real estate industries. By the 1970s, "about 80 percent of the population on Miami Beach was Jewish" (Zerivitz 2019). Many of these original Jewish residents of Miami Beach moved north to Broward and Palm Beach counties but were replaced by a new influx of Jewish residents from Latin America, Russia, and Israel (Heilman 2014).

A housing boom was reignited in the late 1960s, peaking in 1972 with the construction of new suburbs and condominiums. However, the next decade saw a decline in both the reputation and the attractiveness of South Florida. In 1980 about 125,000 Marielitos landed in the United States with most staying in Miami. The Marielitos were sometimes dissidents and sometimes people formerly incarcerated in prisons or treated in mental hospitals in Cuba who departed from the port town of Mariel. Further, between 1977 and 1981 some 60,000 Haitians fled the oppressive regime of Francois Duvalier, also concentrating in Miami. Drug cartels operated in Miami, particularly connected with the Medellin Cartel in Columbia as Miami became known as the "drug capital of the world" (Farzad 2018). Time magazine summarized this disenchantment with a cover story titled "Paradise Lost," leading with an article about South Florida being "hit by a hurricane of crime, drugs and refugees" (Kelly 1981).

Miami's reputation recovered by the beginning of the twenty-first century, and another housing boom ensued until the housing bubble of 2008. This construction expansion was somewhat different from that of previous decades in that it was partly tied to the place of Miami as an international banking center, with particularly strong ties to Latin America. Along with international banking in Miami came drug-trafficking and money-laundering, at the time constituting Florida's biggest industry (Nijman 2011). The once drug-based money was converted into real estate and construction, and eventually to a more diversified economy.

Miami became a destination for migrants fleeing oppressive regimes in Latin American and the Caribbean and became a safe harbor for funds from these countries. The Florida Realtors reported that foreign buyers purchased almost $23 billion in existing single-family housing in 2018 (Florida Realtors 2018). More than half of these purchases were concentrated in the Miami-Fort Lauderdale-West Palm Beach MSA and the primary origin of buyers was Latin America or the Caribbean. Foreign buyers tended to pay more for housing than domestic buyers, and about 70% of the purchases were "all cash" rather than requiring mortgage financing.

Most of the buyers purchasing property *for investment* rather than owner-occupancy also chose property in the Miami MSA, and of these buyers the largest numbers were from Canada, the United Kingdom, Brazil and Venezuela. The intended use of the housing was for a vacation home or for rental.

Although most of the description of the Miami-Dade-Broward-Palm Beach County region has focused on the cities of Miami and Miami Beach, there are other communities within this region that have had very different development histories and present-day characteristics. For example, Palm Beach County is Florida's wealthiest, and includes several luxurious barrier island communities such as Manalapan, Gulf Stream, Jupiter Inlet Colony and Golf, not to mention Mar-a-Lago, which serves as a "second White House" for U.S. President Donald Trump. Like Miami-Dade and Broward counties, it is ethnically diverse and is composed of numerous communities and neighborhoods that are spatially adjacent and yet ethnically and economically different from one another. For example, while Mangonia Park, Belle Glade, Pahokee, Stacey Street, San Castle, South Bay and Riviera Beach are primarily African-American, Briny Breezes, South Palm Beach, Gold, and Jupiter Inlet Colony are almost entirely white. Hispanic concentrations are found in Pine Air, Gun Club Estates, Royal Palm Estates, Cloud Lake Acacia Villas and Kenmore Estates (Statistical Atlas of Palm Beach County 2018).

Each of these areas has its own development history. Two contrasting examples are Riviera Beach and Jupiter Inlet Colony. Riviera Beach was created when Mattie Spencer Heyser changed the name of Oak Lawn to Riviera in 1893. Charles Newcomb bought the Riviera Hotel in 1901 and sold building lots in the 1920s. In the meantime, other developers including William Taylor and George Currie sold lots to black families, establishing a black middle-class neighborhood (Swan 2019). In addition, squatters from the Bahamas settled Singer Island and were engaged in commercial fishing (Palm Beach County history online 2009a, b). Currently, two-thirds of the population of Riviera Beach is Black or African American. Houses currently listed for sale on Zillow in Riviera Beach range in price from $7500 to $650,000.

Jupiter Inlet colony had a different history. It was developed by Charles Prynne Martyn of Massachusetts, who purchased 86 acres and sold residential building lots to purchasers such as singer Perry Como. In the mid-1960s, the development was sold to Jack Kurtz (Palm Beach County History online 2009a, b). Currently, while schools, amenities and crime get “A” ratings from areavibes.com, the cost of living gets and “F”: in April 2019, the 13 houses for sale there, ranged in asking price from $1.3 to $6 million according to Zillow.

The largest city in Palm Beach County is West Palm Beach, with a population in excess of 100,000. Settlement was enabled by an extension of the Florida East Coast Railway by Henry Flagler, who also built the Royal Poinciana Hotel, the Breakers and Whitehall. Flagler had the original street system laid out in a grid pattern. His intention was for Palm Beach to become a resort area, with West Palm Beach becoming a city that would house the resorts’ workforce. The city was incorporated in 1894, and like other parts of South Florida, experienced a real estate boom in the 1920s, followed by a crash, and then a recovery in the ‘50s and ‘60s when it became one of the fastest growing areas in the nation. There are a variety of neighborhoods and communities in the city, and house prices in 2019 ranged from less than $100,000 to more than $10 million.

Broward County, founded in 1925 from portions of Palm Beach and Dade Counties, is also part of the Miami metropolitan area. The county seat is Fort Lauderdale, founded on the site of a stockade built in 1838. Settlement grew quickly after the completion of the Florida East Coast Railway in 1896 and the city was incorporated in 1911. Like other coastal cities in Florida, Fort Lauderdale experienced a land boom, this one spurred by a Boston developer W. F. Morang who dredged wetlands to create a series of seven “finger islands” named the “Lauderdale Isles” (Geocities.com 2008). The islands were subdivided into building lots, and as in other parts of Florida sold by developers to relatively unwary non-residents. The developer used innovative techniques to interest buyers: in this case providing prospective buyers with sightseeing tours by boat. The pace of development was ended with a serious hurricane in 1926 that did extensive damage. The area was re-platted and began to be resold in the late 1930s through the early 1950s.

Other developers created subdivisions using the similar dredge-and-fill system to create finger islands. These properties were connected to the mainland by road and were valued for their offer of waterfront property (Fig. 4.5). Among these finger island communities are “The Isles” developed in the 1920s by Charles Rodes who had studied the canal system in Venice and wanted to replicate it in Florida (Brasted and Lazarus 2019). Eventually 643.738 kilometers (400 miles) of “canals” were created, earning the city of Fort Lauderdale the nickname of the “Venice of America.”

Fort Lauderdale also became known as the site for spring break and beach parties particularly in the 1960s, a phenomenon that was popularized by the 1960 movie *Where the Boys Are* and has continued through the present (Trischitta 2019; Diaz and Duarte 2015). Tourism remains a major industry for Fort Lauderdale, but the city is also the home to several companies such as AutoNation, Spirit Airlines, and Gulfstream International Airlines.

In sum, the coastal urban areas of South Florida share a common history despite idiosyncrasies in their settlement and growth. Several began as very small settlements when Florida was incorporated into the United States in 1822. Their growth

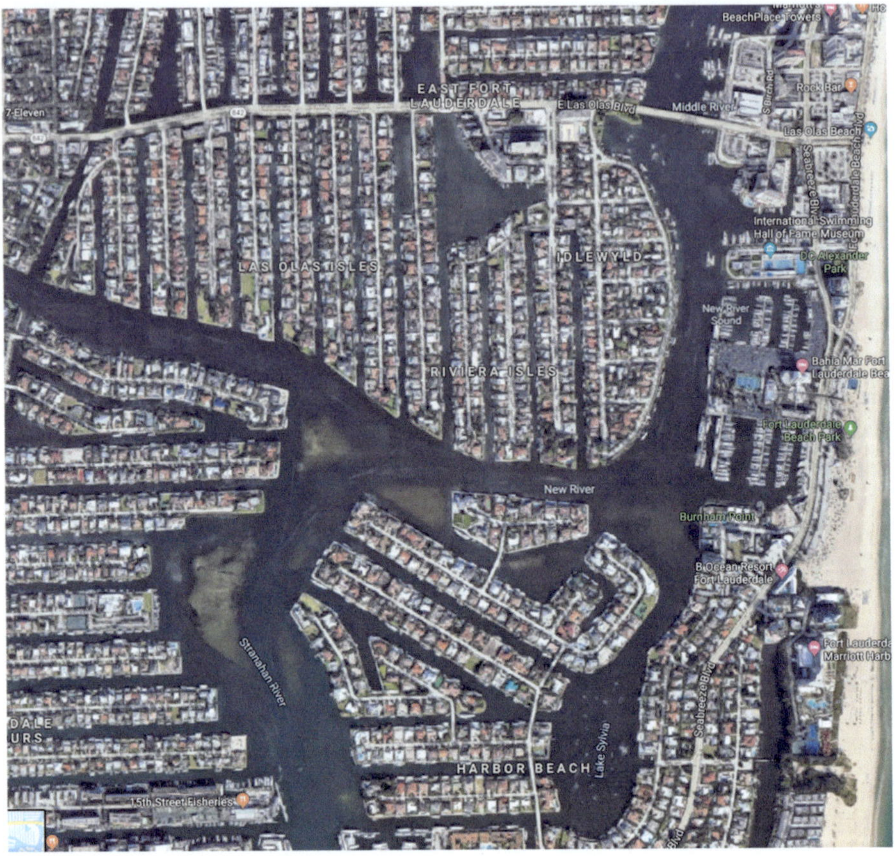

Fig. 4.5 Fort Lauderdale: the "Venice of America". (Source: Google Maps satellite imagery)

was made possible by the draining of wetlands and their connection to the Northeast through Flagler's Florida East Coast Railway or Plant's Savannah, Florida and Western Railroad. All experienced real estate booms in the 1920s and also severe economic slow-downs. Many were used as military training sites and some maintain vestiges of military bases or defense-related investments. All were hampered by hurricanes and sub-tropical disease. Some became foci for organized crime, particularly around smuggling and drug trade. All experienced some form of real estate speculation, and many have large numbers of non-resident property owners who have purchased homes as vacation or rental properties or for speculation. Most of all, these areas share a common location in what has been considered a subtropical paradise, with the benefits and risks that this implies.

4.4 Climate Change in South Florida

The South Florida study region contains four formal climate types according to the classic Køppen classification system. These are a humid subtropical climate for the northern part of the study area, and a combination of tropical savanna, monsoon, and rainforest climates for the southern part. Miami exemplifies the tropical monsoon climate with hot, humid summers, short warm winters and a winter dry season. The Tampa Bay area has a humid subtropical climate with a mild dry season in the winter and a hot, wet season in the summer. The Tampa region experiences high numbers of thunderstorms on summer afternoons. As noted, climate has been one of the major attractions of Florida in attracting tourism and residential development.

But as the global climate is changing, so too is the climate of Florida along with its sea level. Current projections forecast a sea-level rise of between 35.56 and 66.04 centimeters (14 and 26 inches) in the next 40 years (Huber et al. 2017). Floridians can confront this threat to low-lying coastal areas in several ways. The first is by participating with others to reduce the impacts of human activity on the atmosphere, to slow the pace of climate change and associated sea-level rise. This strategy would include support for reductions in carbon emissions, technologies that capture carbon, or other technologies that would re-radiate solar energy into space. These actions taken locally and globally would address the basic problem of human-induced climate change and the impacts of sea-level rise throughout the world would eventually be mitigated.

The second response to climate change and sea-level rise that South Floridians can engage in is a process of "defending" their current settlements. Defense is the traditional response of government or property developers in dealing with unwanted flows of water: dredge, elevate property or streets, dig canals, construct sea barriers, drain the land with pumps, and build levees to change the flow of water. For example, Miami Beach is investing $500 million in raising street levels and installing pumps (Harris 2018). Unfortunately, this strategy has historically not only frequently failed, but also has resulted in degraded coastlines and ecological damage, while providing only a relatively short-term solution (Huber et al. 2017).

A third strategy is "retreat" or community abandonment—that is, moving from coastal areas to higher elevations inland—as a population adjusts to the new reality of more incursions from the sea. Several studies have suggested such a strategy as inevitable in the face of inaction to combat the underlying climate change problem (Turrentine 2018; Loria 2018). However, the notion that trillions of dollars of property will simply be lost to the sea is one that most people find difficult to understand or accept. While "retreat" may become a necessity, it is not a strategy that has yet been adopted on a community scale.

A fourth strategy, the intermediate strategy of adjustment, is more palatable. Adjustment means learning to live with saltwater. This can mean changes in land use, such as the restoration of wetlands, the replanting of mangrove forests, the reconstruction of reefs, and the rezoning and redevelopment of land with the expectation that some of the buffer lands will be routinely flooded (Loria 2018). In a project called "salty urbanism," an interdisciplinary team explored more radical adaptations to rising seas in North Beach Village of Fort Lauderdale including amphibious structures that float on a buoyant base during floods (Fig. 4.6), the development of "high tide gardens" using salt tolerant plants and pervious paving, and the use of water-based transportation (Huber et al. 2017).

Do Floridians recognize the climate-related risks to the state they call home? The short answer to this question is both "yes" and "no." Despite the existence of thousands of studies outlining the impacts of climate change in Florida and threat of sea-level rise, debates over the existence of climate change and sea-level rise still continue in Florida (Chassignet et al. 2017; Korten 2015). A 2015 survey found that overall 57% of the residents of Florida say they are at least "somewhat worried about global warming." When responses were broken down by county however, the mean percentage who were at least somewhat concerned varied from less than 50%, particularly in the panhandle and northern Florida, to more than 70% in the rest of the state (Howe et al. 2015).

Global climate change threatens the very basis of what makes Florida attractive to tourists, prospective home buyers and investors. A publication by the Environmental Protection Agency (2016) outlined a series of negative impacts of ongoing climate change. These include rising sea-levels, eroding beaches, coastal flooding, more intense tropical storms, damage to corals because of rising temperatures and increased ocean acidity, a threat to water resources from damage to the Everglades, warmer summer temperatures reducing agricultural crop yields, and an increase in the number of hot summer days endangering the health of vulnerable populations. In addition to these threats, climate change could mean more frequent algae blooms such as the red tide outbreak in 2018, an increase in mosquito-borne diseases such as dengue and Zika, and the spread of invasive species. Given the dependence of the Florida economy on a positive image to promote the combination of tourism and real estate, these effects are serious indeed.

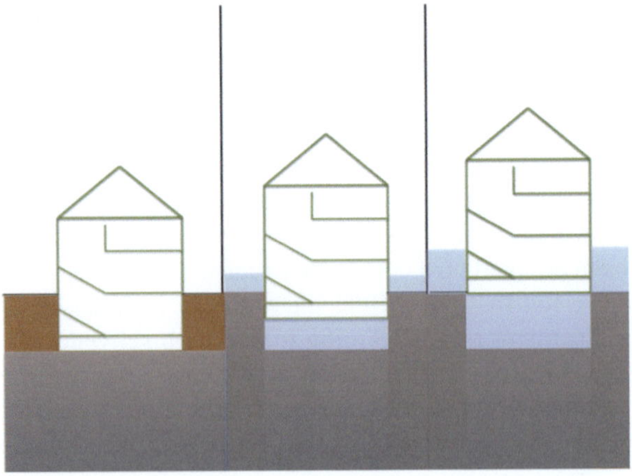

Fig. 4.6 Amphibious structures

How has the state government responded? Florida's previous governor, Rick Scott, was reported to have banned the use of the phrases "global warming" and "climate change" from state documents. The former governor said that he was "not convinced" that there was "man-made climate change" (Schlanger 2015). Scott was subsequently elected to the U.S. Senate, taking office in January 2019. His successor, Ron DeSantis, has taken a neutral position: DeSantis said he did not consider himself a "climate change denier," but he also said he did not want to be labeled a "climate change believer" (Ceballos 2019).

Given the significance of climate for the economic well-being of the region and the varying response to the issue of climate change to date, we consider South Florida an ideal laboratory for the assessment of climate change attitude and policy. While historical sea-level rise in Florida has been a function of a variety of factors, there is no question that a significant rise in ocean temperature and the melting of some of the earth's ice sheets and glaciers, expanding the volume of ocean waters, are an existential threat to this low-lying coastal state. The region is highly dependent on a national and international perception that it has an ideal climate and provides a kind of paradise for retirement or second homes. Many residents seem to be aware of and also concerned about the threats of climate change, and yet at the state level, government leaders have resisted taking action or even recognizing the reality of sea-level rise and climate change. Further, the population has been transient with many recent in-migrants from elsewhere in the United States or immigrants who are unaccustomed to dealing with the local environment. It is against this backdrop that the 2018 survey of residents of coastal postal codes in South Florida was conducted.

References

Arsenault R (1996) St. Petersburg and the Florida dream. University Press of Florida, Gainesville

Artavia D (2018) Why key west endures as a gay paradise. In: Advocate, January 9, 2018. Available at: https://www.advocate.com/travel/2018/1/09/why-key-west-endures-gay-paradise. Accessed 9 Jan 2018

Baker D (2018) The housing bubble and the great recession: ten years later. In: CEPR Center for Economic and Policy Research. Available at: http://cepr.net/images/stories/reports/housing-bubble-2018-09.pdf. Accessed 10 Oct 2018

Bernard E, Hobe D (1983) Lies that came true. Anna Publishing, Inc., Ocoee

Bostick K, Johnson S, Main M (2018) Florida's geological history. In: Department of Wildlife Ecology and Conservation, UF/IFAS Extension, WEC 189. Original publication date April 2005. Revised November 2018. Available at: https://edis.ifas.ufl.edu/uw208. Accessed 14 Jan 2019

Brasted R, Lazarus S (2019) Five great neighborhoods in Fort Lauderdale. In: greatamericancountry.com. Available at: https://www.greatamericancountry.com/places/local-life/great-neighborhoods-in-fort-lauderdale. Accessed 19 Jan 2019

Ceballos A (2019) Gov. Ron DeSantis walks careful line on climate change stance. In: Naples Daily News. Available at: https://www.naplesnews.com/story/news/politics/2019/01/11/gov-desantis-walks-careful-line-climate-change-stance/2547126002/. Accessed 11 Jan 2019

Cedar Lake Ventures (2018) Statistical atlas of Palm Beach County, Florida. In: StatisticalAtlas.com. Available at: https://statisticalatlas.com/county/Florida/Palm-Beach-County/Overview. Accessed 24 Sept 2018

Chassignet E, Jones J, Misra V, Obeysekera J (eds) (2017) Florida's climate: changes, variations & impacts. Florida Climate Institute, Gainesville. https://doi.org/10.17125/fci2017

Cumming JB (2006) A brief Florida real estate history. Appraisal Institute of West Coast Florida, Tampa

De Lea B (2019) US populations growing faster in lower-tax states. In: Fox Business. Available at: https://www.foxbusiness.com/economy/us-populations-growing-faster-in-lower-tax-states. Accessed 6 Jan 2019

Derr M (1989) Some kind of paradise: a Chronicle of man and the land in Florida. William Morrow and Company, New York

Diaz J, Corbellini Duarte B (2015) 55 years after "Where the Boys Are," spring breakers still flock to Fort Lauderdale. In: Sun Sentinel. Available at: https://www.sun-sentinel.com/features/fl-spring-break-where-the-boys-are-20150305-story.html. Accessed 8 Jan 2019

Dodrill D (1983) Selling the dream: the Gulf American corporation and the building of cape coral, Florida. University of Alabama Press, Tuscaloosa/London

EPA: United States Environmental Protection Agency (2016) What climate change means for Florida. EPA 430-F-16-011

Farzad R (2018) Hotel Scarface: where cocaine cowboys partied and plotted to control Miami. Berkley Press/Penguin-Randomhouse, New York

Fisk C (2019) I: ClimateStations.com. Available at: https://www.climatestations.com/miami-fl/. Accessed 7 Apr 2019

Florida Realtors/National Association of Realtors (2018) 2018 Profile of International Residential Real Estate Activity in Florida. Available at: https://www.nar.realtor/sites/default/files/documents/2018-profile-of-international-residential-real-estate-in-florida-10-25-2018.pdf . Accessed 8 Jan 2019

Frazer W, Guthrie JJ Jr (1995) The Florida land boom: speculation, money and the banks. Quorum Books, Westport

Geocities.com (2008) History of the Seven Isles. Available at: https://web.archive.org/web/20091027113149/http://geocities.com/sevenisles2001/history/history.html. Accessed 14 Mar 2018

Foster, MS (2000) Castles in the Sand: the life and times of Carl Graham Fisher. University Press of Florida, Gainesville

Gridics Staff (2016) Visualizing the booms and busts of Miami development through history. In: Gridics.com. Available at: https://www.gridics.com/news/visualizing-booms-busts-miami-development-through-history. Accessed 26 Sept 2018

Grunwald M (2007) The swamp: the Everglades, Florida, and the politics of paradise. Simon and Schuster, New York

Grunwald M (2017) The boomtown that shouldn't exist. In: Politico.com. Available at: https://www.politico.com/magazine/story/2017/10/20/fastest-growing-city-america-florida-cape-coral-215724. Accessed 4 Mar 2018

Harris A (2018) Miami Beach's future is 'uncertain', experts say, but sea pumps are a good start. In: Miami Herald. Available at: https://www.miamiherald.com/news/local/community/miami-dade/miami-beach/article209328849.html. Accessed 28 Apr 2018

Heilman U (2014) Florida: It's not just for old Jews anymore. In: Jewish Telegraphic Agency: jta.org. Available at: https://www.jta.org/2014/02/11/united-states/florida-its-not-just-for-old-jews-anymore. Accessed 18 May 2018

Historical Society of Palm Beach County (2009a) Juniper Inlet Colony. In: Palm Beach County History online. Available at: http://www.pbchistoryonline.org/page/jupiter-inlet-colony. Accessed 29 July 2018

Historical Society of Palm Beach County (2009b) Riviera and Riviera Beach. In: Palm Beach County History online. Available at: http://www.pbchistoryonline.org/page/riviera-and-riviera-beach. Accessed 29 July 2018

Howe P, Mildenberger M, Marlon J, Leiserowitz A (2015) Geographic variation in opinions on climate change at state and local scales in the USA. Nat Clim Chang 5(6):596–603. https://doi.org/10.1038/nclimate2583

Huber J, Van de Riet K, Sandell J, Scarpa L (2017) Salty urbanism: towards an adaptive coastal design framework to address sea-level rise. Plan J 2(2):389–414. https://doi.org/10.15274/tpj.2017.02.02.06

Indian River County Administration (2019) A brief history of Vero Beach, Sebastian, Fellsmere & Indian river county. In: VeroBeach.com. Available at: https://verobeach.com/vero-beach-community/a-brief-history-of-vero-beach-sebastian-fellsmere-indian-river-county. Accessed 3 Mar 2019

Kelly J (1981) South Florida: trouble in paradise. Time Mag 118(21). Available at: http://content.time.com/time/magazine/article/0,9171,922693,00.html. Accessed 14 Aug 2018

Kerstein R (2001) Politics and growth in twentieth-century Tampa. University Press of Florida, Gainesville

Korten T (2015) In Florida, officials ban term 'climate change'. In: Miami Herald. Available at: https://www.miamiherald.com/news/state/florida/article12983720.html. Accessed 14 Apr 2018

Loria K (2018) Miami is racing against time to keep up with sea-level rise. In: Business Insider: Saving Our World. Available at: https://www.businessinsider.com/miami-floods-sea-level-rise-solutions-2018-4. Accessed 12 Apr 2018

Meindl C (2000) Past perceptions of the great American wetland: Florida's Everglades during the early twentieth century. Environ Hist 5(3):378–395. https://doi.org/10.2307/3985482

Milanich J (1998) Florida Indians and the invasion from Europe. University Press of Florida, Gainesville

Miller M (2017) Florida backroads travel: day trips off the beaten path, 2nd edn. FloridaBackroads-Travel.com

Naughton O, McCormack T, Gill L, Johnston P (2018) Groundwater flood hazards and mechanisms in lowland karst terrains. In: Parise M, Gabrovsek F, Kaufmann G, Ravber N (eds) Advances in karst research: theory, fieldwork and applications. Geol Soc Lond Spec Publ 466(1):397–410. https://doi.org/10.1144/SP466.9

NeighborhoodScout.com (2019) St. Petersburg, FL: Crime rates. Available at: https://www.neighborhoodscout.com/fl/st-petersburg/crime. Accessed 30 Apr 2019

Nicholls S, Crompton J (2017) The effect of rivers, streams and canals on property values. River Res Appl 33(9):1377–1386. https://doi.org/10.1002/rra.3197

Nijman J (2011) Miami: mistress of the Americas. University of Pennsylvania Press, Philadelphia

Ogle M (2003) Key West: history of an island of dreams. University Press of Florida, Gainesville

Parks AM (2015) George Merrick, Son of the South Wind University Press of Florida, Gainesville

Purdy B (2008) Florida's people during the last ice age. University Press of Florida, Gainesville

Reaves G (2007) Fort Myers (Then and now: Florida). Arcadia Publishing, Charleston

Sammons S (2010) The two Henrys: Henry Plant and Henry Flagler and their railroads. Pineapple Press, Sarasota

Schlanger Z (2015) Under Rick Scott, Florida environment officials said to be barred from saying 'climate change'. In: Newsweek. Available at: https://www.newsweek.com/under-rick-scott-florida-environment-officials-allegedly-barred-saying-climate-312301. Accessed 18 Apr 2018

Statistical Atlas of Palm Beach County (2018) Available at https://statisticalatlas.com/county/Florida/Palm-Beach-County/Overview

Swan R (2019) FlashBlack: Riviera Beach. In: @palmbeachpost/Historic Palm Beach blog. Available at: http://historicpalmbeach.blog.palmbeachpost.com/2010/02/19/flashblack-riviera-beach/. Accessed 14 Sept 2019

Trischitta L (2019) Spring break traffic invades the beach. And It'll be almost a month before it wanes. In: South Florida Sun Sentinel. Available at: https://www.sun-sentinel.com/local/broward/fort-lauderdale/fl-ne-lauderdale-spring-break-traffic-20190315-story.html. Accessed 18 Mar 2019

Turrentine J (2018) As waters rise, should seaside communities retreat? In: Natural Resources Defense Council: On Earth. https://www.nrdc.org/onearth/waters-rise-should-seaside-communities-retreat. Accessed 29 June 2018
Williams C (2017a) Hidden history of Fort Myers. The History Press, Charleston
Williams C (2017b) No permit required: developing cape coral in 1958. In: News-Press. Available at: https://www.news-press.com/story/news/local/cape-coral/2017/04/19/cape-coral-1958-development-mangroves-leonard-rosen-cynthia-williams/100584564/. Accessed 22 July 2018
Williams C (2017c) Cape Coral in 1957: Building a city on the water. In: News-Press. Available at: https://www.news-press.com/story/news/local/cape-coral/2017/05/24/cape-coral-1957-building-city-water-rosen-williams/335201001/. Accessed 22 July 2018
Zerivitz M (2019) Miami's Jewish history: a brief history of the Jewish community of greater Miami. In: Greater Miami Jewish Federation. Available at: https://jewishmiami.org/about/federation/miami_jewish_history/. Accessed 14 Nov 2018

Chapter 5
Study Methods and General Findings

Abstract The sample was intentionally limited to 166 South Florida postal codes likely to experience either tidal or storm surge flooding. Qualtrics administered an internet-based survey from October through December of 2018 to approximately 1000 individuals. One-half of the sample also received a map of a part of their own city or county showing floodwater depth associated with a Category 3 hurricane in 2033 from the FloodIQ.com website. Analysis of the results shows that, in general, the respondents believe that the climate is changing and that it is human activity that is driving this pace of climate change. However, most do not believe that their own homes are at risk or that their property values will fall as a result of sea-level rise. Respondents on the Gulf Coast gave similar responses to those on the Atlantic coast, although Gulf Coast respondents are less anxious about sea-level rise or the vulnerability of their own home.

Keywords Qualtrics survey · FloodIQ.com · Flood insurance · Hazard disclosure

Instead of a general population survey, or even a survey using county boundaries to define the study population, this study intentionally focused on residents of smaller areas clearly at risk from tidal or riverine flooding associated with climate change. Using maps depicting projected storm surge flooding associated with a Category 3 hurricane developed by FloodIQ.com, we selected 166 postal codes (zip-codes) in South Florida from which to draw the survey sample.

Not all of the residents in the areas we sampled will suffer flooding to their *own* properties in the next 15 years, either from tidal flooding or from storm surge resulting from hurricanes. However, all of the respondents *live in neighborhoods* that will flood, where surface streets will be impassible, and where community facilities will be affected. The survey population was therefore intentionally limited to those living in areas at especial risk from climate change-related flooding.

Qualtrics administered an internet-based survey to residents of these 166 postal codes during the period from October through December of 2018. This is a convenience sample, with all of the advantages and disadvantages that accompany this

R. Palm, T. Bolsen, *Climate Change and Sea Level Rise in South Florida*, Coastal Research Library 34, https://doi.org/10.1007/978-3-030-32602-9_5

mode of survey administration, including the possibility of sample bias due to the lack of coverage of populations that may not have access to the internet. A quality enhancement strategy was used to eliminate questionable responses (Chang 2013) including removal of inconsistent responses, IP address verification to ensure that respondents were actually from the postal codes specified, and removal of responses from repetitive IP addresses. The survey yielded 971 verified responses from the designated study area that are used in this analysis. Qualtrics solicited panel respondents from each of the counties with zip codes that we had identified, leveraging the "routers" of its partners in order to "select respondents for surveys where respondents are highly likely to qualify" (ESOMAR 2014).

We pre-specified quotas with respect to party and gender: half of the respondents were to be men and the other half women, and that one-third of the respondents were to identify themselves as Republicans, one-third as Independents, and one-third as Democrats. We did this pre-specification because we wanted a sample that represented both men and women, and we especially needed to have sufficient numbers of respondents identified by political party since empirical research has shown its overwhelming influence on climate beliefs, an influence that has only increased over time (Driscoll, 2019). We also required that respondents be at least 18 years of age. We did not ask whether the respondents were U.S. citizens.

We did not limit our sample based on residence type or ownership status. Our respondents therefore represented both renters and homeowners, and residents of single-family detached dwellings as well as multiple family dwellings. We directed several questions to the portion of the sample who were owner-occupiers of single family dwellings because they would have considered several different factors in their location process, including the decision to purchase flood insurance. Among these questions were the reasons that they selected their particular home, and whether or not they were provided with a hazards disclosure.

Homeowners' insurance, which is almost universal, covers such hazards as fire and wind damage. But it does not cover flood damage, and the areas we were sampling were susceptible to such damage either to the individual home or at least to the neighborhood. Flood insurance is required for mortgage loans secured by a federally-backed lender, but since only a small minority of American homeowners at risk from flooding actually purchase this addition to their homeowners' insurance, we also asked whether they had purchased flood insurance.

All of the sample population are at risk if not from the direct impact of flooding such as loss of life or property, then at least from the indirect impacts on the local economy as the area becomes perceived as less desirable. Therefore, the climate change beliefs of all respondents, whether invested heavily in home equity or not, were of interest in this study.

5.1 Experimental Design: Use of a Map of Projected Local Flooding

Our experimental design was intended to identify the impacts, if any, of providing a map detailing storm-surge related flooding associated with a Category 3 hurricane on the Sapir-Simpson scale projected for 15 years from now, given present rates of sea-level rise. Before describing the experiment, it is important to note the way in which these risk maps we used were developed. Although they incorporate estimates of sea-level rise for South Florida, it is also clear that estimation of sea-level rise for a local area is a task fraught with difficulties. As Rahmstorf (2012) has summarized it, local sea-level rise can vary from global trends as a result of multiple processes working simultaneously in different ways including winds driving the ocean water in particular places, the gravitational pull of land ice, and changes in local elevation because of forces including tectonic processes, groundwater extraction, land rebounding from glaciation or soil compaction. The Intergovernmental Panel on Climate Change of the World Meteorological Organization (the IPCC) chapter on sea-level change states that local changes in sea-level rise can vary significantly in different parts of the globe and also details the numerous uncertainties in the ocean/climate models and the resulting problems of calculating precise predictions for local areas (Church et al. 2013).

In Florida, numerous regional studies have undertaken the estimation sea-level rise. For example, a climate science advisory panel estimated of sea-level rise for the Tampa Bay Region using a tool developed by the US Army Corps of Engineers combining present-day observations, global parameters of sea-level rise, and historic local rates of change (Tampa Bay climate science advisory panel 2015). This calculation produces a range of possible future states. In the case of St. Petersburg, Florida, the estimates of sea-level rise for the year 2100 range from an increase of 1–7 feet depending on the NOAA global sea-level rise scenario used.

The maps used in this study incorporated sea-level rise in their portrayals of the extent of likely storm surge from hurricanes (McAlpine and Porter 2018). These were derived from the storm surge models used by the U.S. National Weather Service (a branch of the National Oceanographic and Atmospheric Administration or NOAA). The model runs computer simulations known as SLOSH (Sea Lake and Overland Surges from Hurricanes) to predict how hurricane force winds will drive sea water inland. These models are relatively coarse in scale and cannot by themselves be used to indicate how floodwaters will affect particular plots of land. To translate the SLOSH models into something more usable for individual homesites, Flood IQ.com interpolates a surface by taking a high density sampling the SLOSH data and combining this with detailed elevation measures from the Light Detection and Ranging (LIDAR) digital elevation model. They add a projected sea-level rise as calculated by the US Army Corps of Engineers "High SLR" curve, which is "computed from the modified NRC [National Research Council] Curve III considering both the most recent IPCC projections and the modified NRC projections with the local rate of vertical land movement added" (US Army Corps of Engineers 2019). The "high" level sea-level rise scenario assumes a "high-end, fossil-fuel-intensive, 'business-as-usual' emission scenario" (Sweet et al. 2017). The intended result of the final resolution is the measurement of "the most realistic proportion of the lot inundated by each of the flooding scenarios" (McAlpine and Porter 2018).

One-half of the sample received a map tailored to the specific region in which the respondent lived. Using the FloodIQ.com website (https://floodiq.com), we selected a map scale that would permit the respondents to easily locate their own city block. We used seven maps, one for each of the seven regions we surveyed. For example, one-half of the respondents from Fort Myers were shown a map or projected flooding in Fort Myers associated with a Category 3 hurricane 15 years from now given sea-level rise (Fig. 5.1).

Similarly, one-half of the Fort Lauderdale respondents were presented with a map of the hurricane scenario showing flooding in their local area (Fig. 5.2). In the same way, respondents from the other five areas were provided with maps of their own city or county at approximately the same scale and with the same scenario.

Respondents were *not* given instructions to look up their own home address on the website, but if they did, they would have been able to see very detailed projections about the depth of flood waters in various flooding scenarios. They could choose from such categories as "frequent tidal floods," "highest annual tidal flood," and hurricane Category 1 or 3, for both the year during which the survey was taken and for 2033.

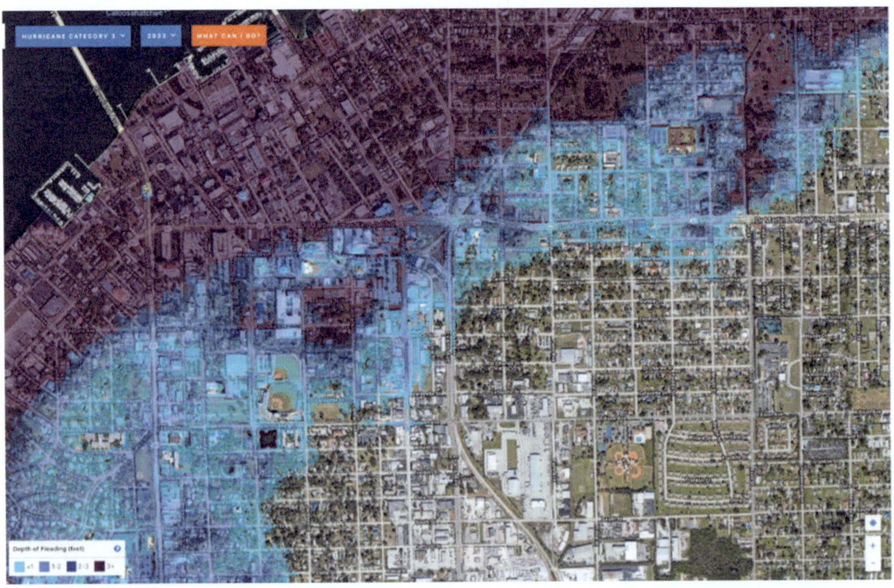

Fig. 5.1 Projected storm surge flooding for Fort Myers. (Source: FloodIQ.com)

Fig. 5.2 Projected storm surge flooding in Ft. Lauderdale. (Source: FloodIQ.com)

We compared the responses of those who had received this map with the responses of individuals assigned to the control group who had not been shown such a map. Our purpose was to see if exposure to a specific and relatively local map of flood risk would shift beliefs about climate change and the local effects of sea-level rise. The questionnaire was structured to test several of the relationships that have been documented in previous studies reviewed in Chap. 2. These studies, undertaken at the national or county scale, showed links between climate change beliefs and such factors as political party, general environmental attitudes, and demographic characteristics.

A full version of the questionnaire is available in Appendix 2. In addition to general demographic questions about race, income, gender, education level, political party identification, and political ideology, we asked questions concerning their general attitudes about hierarchy versus egalitarianism, community versus individual orientation, their trust in science, and their interest in politics. In addition, we asked a battery of six questions to ascertain their general knowledge about science.

After showing half the sample the maps described above, we then posed a series of questions to get their opinions about a series of items, in order to compare those who had seen the maps with those simply answering the questionnaire without this treatment. Topics included:

- their beliefs about the reality of climate change and sea-level rise
- their views on the impacts of sea-level rise on the property values for their neighborhood and their own home

- their support for policy measures that could mitigate the impacts of sea-level rise such as laws regulating property set-backs or the construction of barriers
- their support for laws that could reduce carbon emissions such as direct taxation based on carbon emissions or an increase in the price of fossil fuels
- their beliefs about the extent to which there is a consensus among scientists that humans are having a significant impact on climate change
- their levels of concern about the impacts of climate change for themselves, their families or their communities
- their anxiety about sea-level rise
- beliefs about the efficacy of individuals to slow climate change
- their level of interest in learning more about climate change.
- their beliefs concerning whether or not climate change research is politically motivated
- their opinions as to whether the concept of human-induced climate change is a hoax or conspiracy

5.2 General Findings

We first assessed the characteristics and attitudes of the full set of respondents, whether or not they had been exposed to a map portraying flooding associated with a Category 3 hurricane.

5.2.1 Demographic and Geographic Characteristics of the Respondents

Although the respondents were nearly equally divided between male and female, about 74% of the Democratic respondents were female while and only 41% and 38% of the Republican and Independent respondents respectively were female. Respondents ranged in age from 18 to 92, with both a mean and the median of just over 50 years old. Republican respondents were on average about 3 years older than Democrats and reported higher estimated family household incomes.

In order to benchmark the scientific literacy of our respondents, we posed six limited-choice questions of varying difficulty. These included: which gas makes up most of the Earth's atmosphere (multiple choice), whether seasons are caused by the Earth being tilted on its axis (true/false), whether lasers work by focusing sound waves (true/false), how long does it take for the Earth to go around the Sun (multiple choice), which kind of waves are used to make and receive cellphone calls (multiple choice), and what does a light-year measure (multiple choice). There were no significant differences among the parties in the likelihood of correctly answering any of these questions. The one most frequently missed was the composition of the earth's atmosphere, with most respondents answering "oxygen" instead of the correct answer of "nitrogen." The question most frequently answered correctly was the length of time it takes for the earth to go around the sun (with 74% correctly selecting 1 year as opposed to 1 day or 1 month), and the fact that seasons are caused by the Earth's tilt on its axis (with 73% answering this correctly as "true"). Independents were more likely to answer most of the questions correctly, with 42% of Independents but only 32% of Republicans and 28% of Democrats answering at least five of the six questions correctly. The average respondent answered three of the six questions correctly.

5.2.2 *Residence and Homeownership*

Most of the respondents, regardless of party, had lived in Florida for more than 5 years, with the median residency in the state at 20 years. More than half of the respondents overall lived in single-family detached dwellings, and Republicans were more likely than either Independents or Democrats to occupy single family dwellings (65% of Republicans, 52% of Independents and 49% of Democrats). Similarly, Republicans were more likely to own their own home regardless of dwelling type (77%) than either Independents (51%) or Democrats (38%).

Homeowners were asked about the factors that were important to them in their purchase decision (Table 5.1). Safety from crime was the most important factor, while quality of the local public schools ranked last. The fact that the average age of the respondents who were owners of single-family residents was 51.2 may at least have partially accounted for the relatively small impact reported of school quality on the purchase decision.

We asked homeowners about the risk of flooding to the property or neighborhood, and whether such information had been disclosed to them as part of the purchase process. In order to interpret the responses, we should note that Florida is one of the 21 states with *no statewide law* requiring disclosure that the house has previously flooded (Scata 2018). However, as is the case for the rest of the nation, sellers must disclose "any facts or conditions they know about that materially affect the

Table 5.1 Factors important in the home purchase decision

Factors important to home purchase	Average score out of 7
Safety from crime in neighborhood	6.02
Safety from flooding	5.27
Proximity to entertainment	4.84
Proximity to the coast	4.70
Proximity to work	4.46
Quality of the local public schools	4.00

value of the property" including water leaks, roof defects, mold, termites, problems with wiring, or disputes about the property's boundary (Smith 2019).

Disclosure of flooding takes place in two ways: the first is through the mandatory purchase of flood insurance for homes in a designated Federal Emergency Management Agency (FEMA) 100-year floodplain that have a federally-insured mortgage loan. If a property is bought in cash or through a private loan, no such mandatory flood insurance is required.

The second mode of disclosure has been developed by county governments. For example, as of 2014, Broward County provides a detailed map of FEMA flood zone designations for individual properties with advice as to elevation and insurance requirements. In 2015, Pinellas county officials, in association with representatives from real estate agents, the insurance industry and the mortgage-lenders, developed both a brochure with maps of storm surge and flood-prone areas and a plan for public outreach. Miami-Dade County has a similar flood zone map and also requires disclosure by the seller that the property is in either a "special flood hazard" or "coastal high hazard" area.

Despite the existence of these modes of disclosure, there are several reasons why they may have less impact than might be anticipated. The first is that when hazard disclosure is done at the closing along with all of the other paperwork involved in the purchase transaction, the hazard disclosure is often less salient to the buyers (Palm and Hodgson 1992). In addition, disclosures sometimes come with equivocal language that reduces the impact of the disclosure by stating essentially that all property is subject to the same risk, such as that of Broward County's materials, that "all areas are susceptible to flooding, although to varying degrees." Given these problems, it is perhaps not surprising that despite the fact that all of our respondents live in areas susceptible to serious storm surge or riverine flooding, fewer than half remembered a disclosure by a real estate agent or mortgage lender. Regardless of political party, only about 42% recalled if "anyone, such as the real estate agent or the mortgage lender, discussing flooding during the process of purchasing the home." Only a minority of the respondents (37%), again regardless of party, reported that they had purchased a separate flood insurance policy, even though this is required of purchasers who have a federally-insured mortgage loan. Nationally, about 4% of American households have flood insurance policies, with about 10% of all households living in either the 100- or the 500-year floodplain (Montano and Savitt 2018).

5.2.3 *Climate Change Beliefs*

The vast majority of respondents, Republican or Democrat, indicate that they believe that climate change is taking place. The average respondent indicated that climate change "very likely is happening." One-third of the Republicans but only 7% of the Democrats either believed that climate change was not happening or were not sure (Fig. 5.3). This partisan difference was consistent for many of the questions

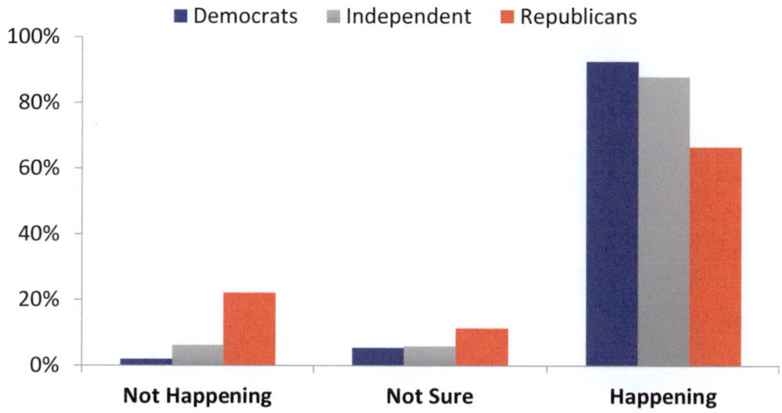

Fig. 5.3 Belief that climate change is taking place

we posed. In addition, Republicans in Florida were more likely to agree that climate change is taking place than those nationally: although 66% of the South Florida Republicans believe climate change is happening, only 52% expressed such a view in a 2016 survey conducted by Yale that drew a nationally representative sample in the U.S. (Mildenberger et al. 2017).

Even though it should be apparent to this particular population that the sea level has risen and that their own homes and communities are increasingly susceptible to high tides and coastal flooding, the same partisan difference of opinion exists when people are asked whether they believe the sea level is rising. The modal response from Republicans was that the sea level "probably" is rising, while the modal response for both Independents and Democrats is that it "definitely" is rising. Some 37% of the Republican respondents were either not sure or attested that the sea level is *not* rising (Fig. 5.4).

It might be expected that even those who doubt the existence of sea-level rise would be concerned about its possible impact on property values, particularly given news stories such as "the trillion-dollar coastal property bubble is ready to burst," containing warnings that "at some point, the property becomes worthless" (Romm 2018). This concern might especially affect owner-occupiers, since home equity often makes up a major portion of a household's wealth. Yet, only a minority of the homeowners in our sample believe that *their own homes* are susceptible to sea-level rise. Again, perceived susceptibility of the home, neighborhood or property values to sea-level rise followed party lines (Fig. 5.5). Only 36% of Republicans but 43% of Democrats believe that their own home is susceptible to sea-level rise. Although there are clear partisan differences, a majority of respondents, regardless of party, *do not believe or are unsure* about whether their own homes are susceptible to sea-level rise.

Of course, if a home is not susceptible to sea-level rise, its market value should also not be affected by climate change. On the other hand, if other property in the neighborhood is flooded or if streets are flooded, one might think that this would

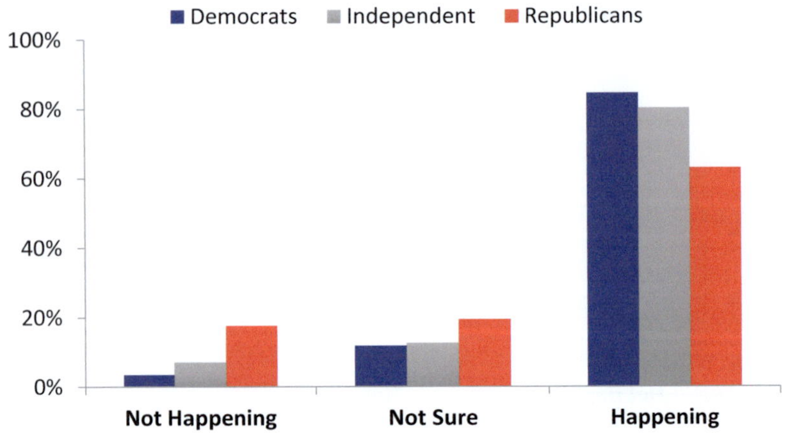

Fig. 5.4 Percentage who believe that sea-level rise is taking place

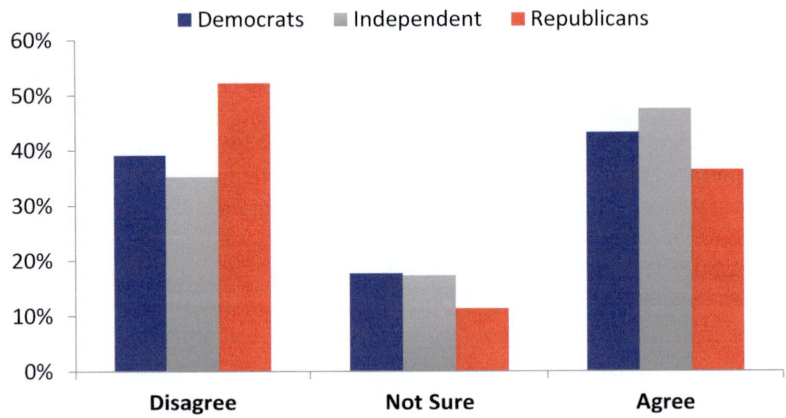

Fig. 5.5 Belief that own home is susceptible to sea-level rise

have an effect on property values. Despite this expectation, the modal response of Republican homeowners was that even if the sea-level rises, their *own* property value would be affected "not at all." The modal response for Democrats for this question was "somewhat." Both groups conceded, however, that if the sea-level rises, it could reduce property values in their *community*. The modal response for both Republican and Democratic homeowners was that there would be a "great deal" of impact on property values, the response given for about 30% of the Republicans and 40% of the Democrats. Only 10% of the Republicans and 1.6% of the Democrats said that there would be no reduction in property values ("not at all") in their communities in the event of sea-level rise (Fig. 5.6).

Partisan politics even divides opinion as to whether or not sea-level rise is related to climate change. Although most scientists believe there is a clear relationship between the increased amount of water in the ocean caused by the melting of polar

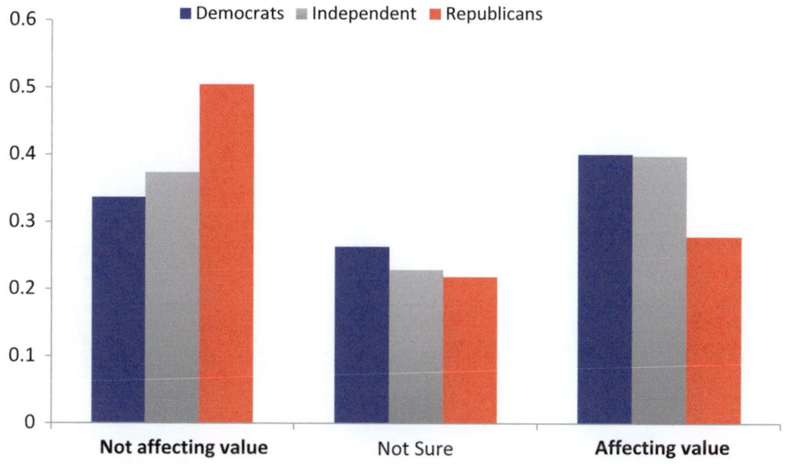

Fig. 5.6 Sea-level rise will affect property value of own home

ice and the expansion of ocean volume associated with warming waters, there has been some opinion expressed that the sea-level is not rising at all, but instead the land is sinking or there are tectonic forces increasing the amount of rocky material in the ocean beds. In fact, one congressman testified that it is the falling of rock into the ocean that is causing apparent sea-level rise (Waldman 2018). Politics therefore divides people when they are asked about the connection between climate change and sea-level rise. Democrats overwhelmingly said sea-level rise is "definitely" related to climate change, while Republicans responded either that they were "not sure," or that it "probably is" related to climate change (Fig. 5.7).

Beliefs about the impacts of climate change on the likelihood of more extreme storms and hurricanes also followed partisan lines. When asked whether climate change is responsible for causing "more extreme storms and hurricanes," the modal response for Republicans was "not at all," while the modal response for Democrats was "a great deal."

5.2.4 Other Responses to Sea-Level Rise and Climate Change

The survey included several other questions probing the emotional responses to the threats of climate change and sea-level rise, beliefs in personal efficacy in dealing with climate change, and attitudes towards climate change research generally.

When asked on a 1–7 scale from "extremely unconcerned" to "extremely concerned," "how concerned are you about the effects that climate change will have on you, your family, and your community," the modal response was "somewhat concerned." Again, this varied by party affiliation with Republicans and Independents most likely to say "somewhat concerned," and Democrats most likely to say "very concerned." For Republicans, 34% said they were unconcerned, while only 11% of Democrats were unconcerned.

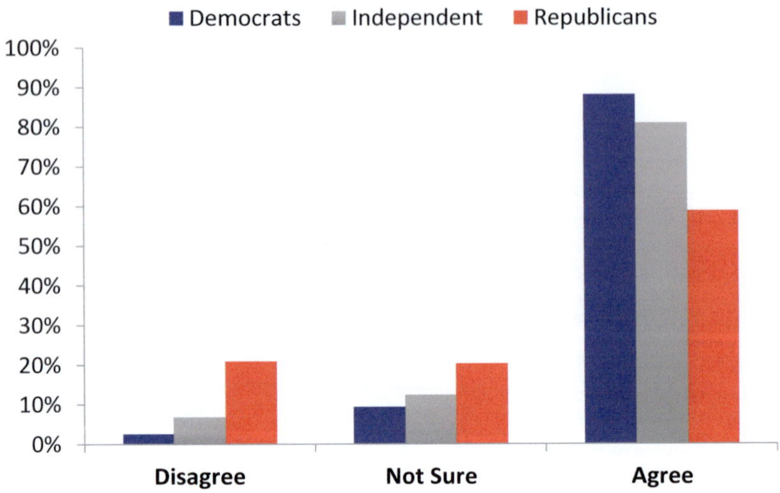

Fig. 5.7 Belief that sea-level rise is caused by climate change

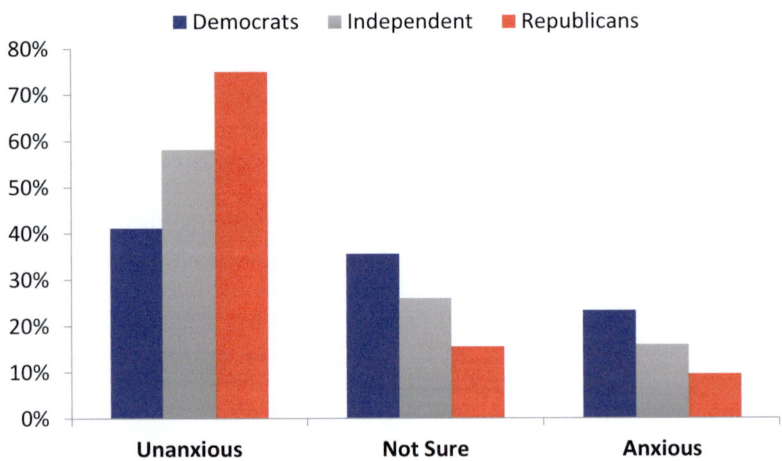

Fig. 5.8 Anxiety about sea-level rise by political party

A similar question seeking information about the emotional response to sea-level rise asked "as you think about sea-level rise, how much anxiety do you feel?" Responses on a 5-point scale ranged from "none at all" to "a great deal." The modal response overall was "none at all." For Republicans, 75% of respondents were not anxious, but only 41% of Democrats were not anxious (Fig. 5.8).

People also vary in their beliefs about their own efficacy to slow climate change and also the responsibility of the government. Respondents were asked to respond on a 7-point scale to whether or not they agreed with this statement: "I believe my actions have an influence on global warming and climate change." What is being

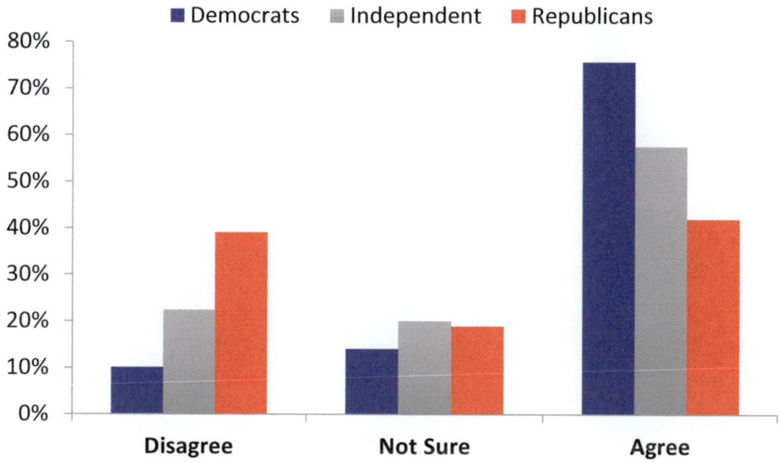

Fig. 5.9 Belief in effectiveness of personal action to influence climate change by political party

measured here is two ideas: first, whether *human beings can* affect global warming and climate change, and second, whether *they* as individuals have the power to bring about such an effect. The modal response to this question was "slightly agree" for all political parties, but with a strong skew towards "strongly disagree" by Republicans (Fig. 5.9). While 39% of Republicans disagreed with this statement, only 10% of Democrats disagreed.

Respondents were asked about the responsibility of government for dealing with climate change on a 7-point scale, and whether they agreed or disagreed with this statement: "It is governments, not citizens, that are primarily responsible for dealing with climate change." Again, this question assumes that the respondents believe that humans can influence climate change, and then asks whether the responsibility for such action is individual or collective. Republicans and Independents were more likely to disagree with this statement (58% and 52% respectively) than were Democrats (44.6%): in other words, Democrats were more likely to accept the notion that people *can* slow climate change and that it is the government's responsibility to do so.

Research has sought to identify whether people are willing to take further action on climate change. One mode of "action" is simply to become more informed about the subject. We asked whether the respondents were interested in learning more "about the possible effects of sea-level rise." The modal response was "somewhat interested," (30% overall) but with Democrats more likely to be "very or extremely interested" (44%) than Republicans (22%) or Independents (24%).

As described in Chap. 2, belief that there is a consensus among scientists about climate change has been posited as a "gateway belief," increasing both belief in climate change itself and also support for climate policy (van der Linden et al. 2015). We therefore posed this question: "Do you agree with the following statement: 'Most scientists think climate change is happening primarily as a result of human activities?'" While most of the respondents "strongly agree" or at least

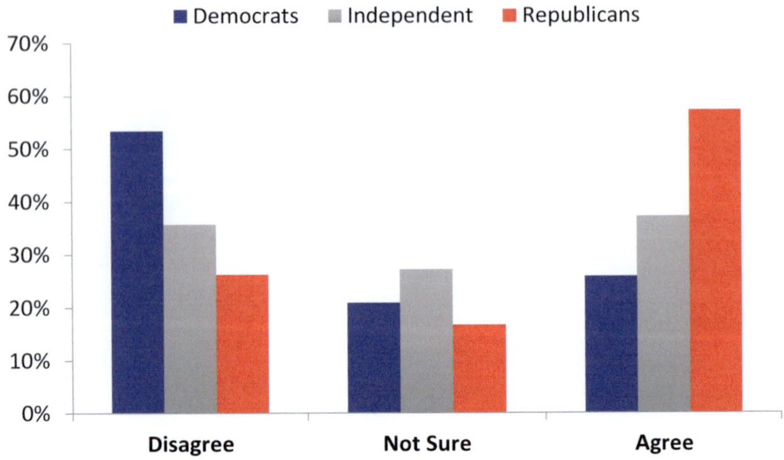

Fig. 5.10 Belief that climate science research is "politically motivated"

"moderately agree" with this statement, there were significant numbers of both Independents and Republicans who either "disagreed" or "neither disagreed nor agreed" (39.7% of Republicans and 27.2% of Independents).

Finally, we posed two questions that are based on statements that have been made by those who doubt the existence of climate change: that scientific research on climate change is politically motivated, and that "the idea that climate change is primarily due to human activities is a hoax or a conspiracy." On the issue of the political motivation of scientific research, there was a wide range of opinion, although again mostly sorted by partisan affiliation (Fig. 5.10). The modal response was "neither agree nor disagree." However, while 53% of Democrats disagreed with the notion that the research is politically motivated, only 36% of Independents and 26% of Republicans shared their view.

The vast majority of the respondents do not believe that the idea of anthropogenic climate change is a conspiracy or hoax. For the overall sample, 61% disagreed with the statement, with 74% of Democrats disagreeing. Among Republicans, however, 35.5% did agree with the statement that human-induced climate change is a hoax. For those who said that they "strongly agree" that the idea is a hoax," 35% also identified themselves as "very conservative" and another 28% said they were "mostly conservative."

5.3 Regional Effects

In addition to the effects of political party on response, we also wanted to investigate any impacts of location on perceived risk of climate change and sea-level rise. Although South Florida is itself considered a region, it can also be divided into the west or Gulf coast as opposed to the east or Atlantic coast. Do these different coastal areas also respond differently to sea-level rise?

Why might there be differences in response to sea-level rise and climate change on the two coasts? Both the nature of the residential settlement and differences in climate could account for some differences. It has been posited that the two coasts differ culturally partly because the Gulf Coast of Florida was settled by midwesterners following Interstate 75 while the east coast was settled by New Yorkers driving down Interstate 95 (Granville 2018). While this highway-oriented migration path may have existed, it has more recently been replaced with a complex domestic and international migration pattern. According to the U.S. Census American Community Survey estimates, the top source of domestic migration to Miami-Dade County on the Atlantic coast was the New York City metropolitan area, followed by Boston and Chicago between 2012 and 2016. For Tampa, on the Gulf Coast, domestic migrants came from Chicago and Houston, but also large numbers were from New York. The fact that migration patterns have converged suggests that there is likely very little cultural difference at present. And, even if there are some small cultural differences, these would not predict variability in the way sea-level rise or climate change is evaluated by the local population.

A more important distinction between the regions would be the differences in geographic settings and therefore the specific local risks. Is the climate, or the risk of hurricanes different from one coast to the other? The simple answer to that question is "not very much." As has been noted earlier, most of the state is categorized as humid subtropical with humid and hot summers and cooler or mild winters. Rainfall is concentrated in the summer months with relatively dry fall and winter seasons. The southeast part of the state including the Florida Keys are classified as tropical wet-and-dry, with a similar variation in annual rainfall but higher winter temperatures than those further north in Florida. The Gulf coast is historically at less risk from tropical storm force winds, being classified as "high wind risk" as opposed to "extreme wind risk" but is at greater risk from storm surge associated with hurricanes regardless of category (Emrich et al. 2014). In short, hurricanes are a hazard on both coasts (Fig. 5.11).

Because of local conditions including ocean currents and subsidence, the trend in sea-level rise is somewhat different on the Gulf and Atlantic coasts. For example, Key West is estimated to experience a sea-level rise of 2.42 millimeters per year based on the period from 1913 to 2018, with a total change of 24.13 centimeters (9.5 inches) over the 100-year period (NOAA/National Ocean Service 2018). This figure is similar in Miami Beach (2.39 mm per year). Fort Myers has experienced a higher rate of sea-level rise, of 3.11 mm per year just over 30.48 centimeters (one foot) rise in 100 years. Clearwater Beach just north of St. Petersburg, has experienced an even higher rate of change with 3.68 millimeters per year or a total of 36.83 centimeters (14.5 inches) in 100 years. In sum, the differences are relatively small, but the Gulf Coast residents have experienced a somewhat larger increase in sea-level rise than those on the Atlantic Coast.

In our 2018 survey, about 62% of the respondents lived on the Gulf Coast, including the Tampa-Saint Petersburg-Clearwater area, and Fort Myers. The remainder, from Miami-Dade County, Fort Lauderdale, West Palm Beach and Palm Beach, and Vero Beach, were classified as East Coast. The samples varied little in demographic

Fig. 5.11 100 years of storm tracks. (Source: Photo from National Oceanic and Atmospheric Administration)

characteristics except for average age: the average age of the East Coast respondents was 50.8 years while that on the Gulf Coast was 55.5 years. Comparing the populations using chi square tests, there were no statistically significant differences between the two populations with respect to gender, ideology, party affiliation, or attitude with respect to the balance between prioritizing the environment vs. the economy. They were also not different in their perceptions of whether or not sea-level rise is taking place, property values in their city would be affected if there were sea-level rise, climate change taking place, climate change affecting sea-level rise, trust in science, or desire to learn more about climate change. They also did not differ in attitudes such as whether we have pushed too far with respect to equal rights, that we would be better off if the government did not try to solve everyone's problems, their general interest in politics, their knowledge of science, their perceived personal efficacy in affecting climate change, or their attribution of responsibility to individuals or the government. They equally supported or opposed such mitigation policies as zoning changes to increase set-backs from the shore, government buy-backs of property that was susceptible to flooding, and the adoption of a carbon tax. Finally, there were no regional distinctions in whether or not they believed that scientific research on climate change is politically motivated, that most scientists agree that the main cause of climate change is human action, and that the idea of climate change caused by human activities is a hoax or conspiracy.

There were regional differences in four dimensions, however. These included beliefs about the susceptibility of their own home to sea-level rise, beliefs about their own property values being reduced if sea-level rise occurs, support for two of the mitigation measures, and levels of concern and anxiety about climate change and sea-level rise.

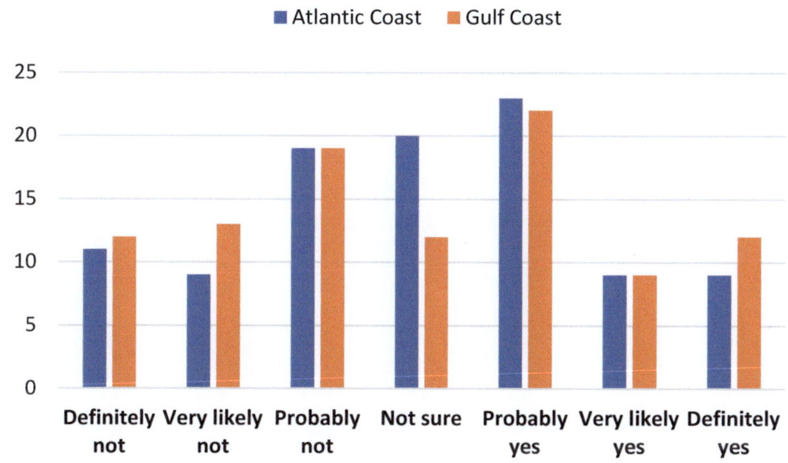

Fig. 5.12 Percentage responses for perceived home vulnerability by region

Gulf Coast residents were overall somewhat less likely to believe that their own home is susceptible to sea-level rise than were respondents from the Atlantic Coast and were also less likely to be unsure about this vulnerability (Fig. 5.12). This difference in perceived susceptibility exists despite the fact that the sea-level is actually rising slightly faster on the Gulf Coast than on the Atlantic Coast, and in any case, coastal settlement on both coasts is at an elevation below 4 meters (13.12 feet).

Gulf Coast residents are also less likely to believe that sea-level rise will affect the property value of their home (Fig. 5.13) with the modal response being "not at all," in contrast to the modal Atlantic coast response of "somewhat." The two regions were similar however with respect to percentages who believed their property values were particularly at risk: about one-fifth of the respondents of both areas said they believed that their property values would be reduced "a great deal" in association with sea-level rise.

With respect to mitigation measures, there was little difference between the two regions except for support for the construction of barriers or sea walls and also the imposition of a tax on carbon-based fuels to encourage energy saving. Atlantic coast respondents were more likely to favor the construction of sea walls: 28.5% strongly supported this measure as opposed to 19% of the Gulf Coast respondents. They were also more likely to strongly support increasing the price of fossil fuels with 17% of the Atlantic coast residents but only 10% of the Gulf coast residents expressing this level of support.

The sharpest contrasts were the emotions that sea-level rise and climate change elicit in the two populations. When asked how concerned are you about the effects that climate change will have on you, your family, and your community, Atlantic coast respondents were more likely to express high levels of concern: 22% of Atlantic coast respondents but only 13% of Gulf Coast respondents said they were "extremely concerned." Similarly, when asked how much anxiety is felt about sea-level rise, Gulf Coast respondents were more likely to say none at all or only a little (Fig. 5.14).

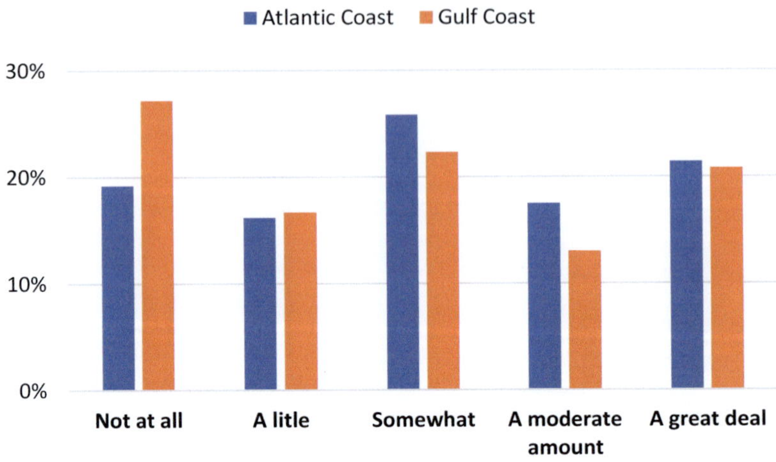

Fig. 5.13 Perceived risk of reduction in property value by region

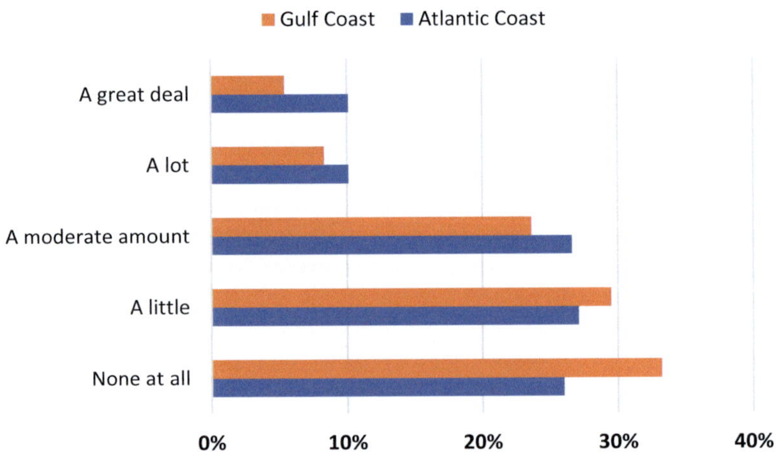

Fig. 5.14 Anxiety about sea-level rise by region

To summarize, residents of the two Florida coasts are similar in most of the dimensions of response to sea-level rise. What differentiates them primarily is their level of confidence that their own home is safe and that they have nothing to worry about. The respondents from the Gulf Coast show far less anxiety about sea-level rise, less concern about climate change in general, and less concern either that their own home is vulnerable or that the value of their property could decrease with sea-level rise. This relative degree of confidence is not justified on the basis of real differences in risk.

References

Chang T-Z (2013) Strategies for improving data reliability for online surveys: a case study. Int J Electron Comm Stud 4(1):131–140

Church J, Clark P, Cazenave A, Gregory J, Jevrejeva S, Levermann A, Merrifield M, Milne G, Nerem R, Nunn P, Payne A, Pfeffer W, Stammer D, Unnikrishnan A (2013) Sea-level change. In: Stocker T, Qin D, Plattner G, Tignor M, Allen S, Boschung J, Nauels A, Xia Y, Bex V, Midgley P (eds) Climate change 2013: the physical science basis. Contribution of working group I to the fifth assessment report of the intergovernmental panel on climate change. Cambridge University Press, Cambridge

Driscoll, D (2019) Assessing sociodemographic predictors of climate change concern, 1994–2016. Soc Sci Q. https://doi.org/10.1111/ssqu.12683

Emrich C, Morath D, Bowser G, Reeves R (2014) Climate-sensitive hazards in Florida: identifying and prioritizing threats to build resilience against climate effects. In: Florida BRACE (Building Resilience Against Climate Effects), Final Report to the Florida Department of Health, Bureau of Epidemiology. Hazards and Vulnerability Research Institute, Columbia. Available at: https://flbrace.org/images/docs/climate-sensitive-hazards-in-florida-final-report.pdf. Accessed 21 June 2018

FloodIQ.com (2018). Available at: https://floodiq.com/

Granville K (2018) East vs. West: How the vibes differ on Florida's two coasts. In: Mansion Global. Available at: https://www.mansionglobal.com/articles/east-vs-west-how-the-vibes-differ-on-florida-s-two-coasts-114327. Accessed 2 Dec 2018

McAlpine S, Porter J (2018) Estimating recent local impacts of sea-level rise on current real-estate losses: a housing market case study in Miami-Dade, Florida. Popul Res Policy Rev 37(6):871–895. https://doi.org/10.1007/s11113-018-9473-5

Mildenberger M, Marlon J, Howe P, Leiserowitz A (2017) The spatial distribution of republican and democratic climate opinions as state and local scales. Clim Chang 145(3–4):539–548. https://doi.org/10.1007/s10584-017-2103-0

Montano S, Savitt A (2018) The cost of flood insurance is a price worth paying. In: Citylab.com. Available at: https://www.citylab.com/environment/2018/08/the-cost-of-flood-insurance-is-a-price-worth-paying/566363/. Accessed 9 Sept 2018

NOAA National Oceanic and Atmospheric Administration 2018. Tides and currents. National Ocean Service. Available at: https://tidesandcurrents.noaa.gov/sltrends/. Accessed 14 Nov 2018

Palm R, Hodgson M (1992) Earthquake insurance: mandated disclosure and homeowner response in California. Ann Assoc Am Geogr 82(2):207–222. https://doi.org/10.1111/j.1467-8306.1992.tb01905.x

Qualtrics (2014) ESOMAR 28: 28 Questions to help research buyers of online samples. Available at: https://success.qualtrics.com/rs/qualtrics/images/ESOMAR%2028%202014.pdf. Accessed 14 July 2017

Rahmstorf S (2012) Modelling sea-level rise. Nature Education Knowledge 3(10):4

Romm J (2018) The trillion-dollar coastal property bubble is ready to burst, per new study. In: ThinkProgress. Available at: https://thinkprogress.org/rising-seas-hit-u-s-coastal-property-values-a-pricing-signal-from-climate-change-848bf4e7443b/. Accessed 25 Apr 2018

Scata J (2018) Home buyers face stacked deck to learn of past floods. In: National Resources Defense Council. Available at: https://www.nrdc.org/experts/joel-scata/home-buyers-face-stacked-decks-learn-past-floods. Accessed 20 Aug 2018

Smith E (2019) What does a seller of a home have to disclose in Florida? In: The Robertson Firm. Available at: http://robertsonfirm.com/what-does-the-seller-of-a-home-have-to-disclose-in-florida/. Accessed 2 Feb 2019

Sweet W, Kopp R, Weaver C, Obeysekera J, Horton R, Thieler E, Zervas C (2017) Global and regional sea-level rise scenarios for the United States. NOAA Technical Report NOS CO-OPS

083. Available at: https://tidesandcurrents.noaa.gov/publications/techrpt83_Global_and_Regional_SLR_Scenarios_for_the_US_final.pdf. Accessed 16 Aug 2019

United States Army Corps of Engineers (2019) Sea-level change curve calculator (Version 2019.21). Available at: http://corpsmapu.usace.army.mil/rccinfo/slc/slcc_calc.html. Accessed 16 Aug 2019

van der Linden S, Leiserowitz A, Feinberg G, Maibach E (2015) The scientific consensus on climate change as a gateway belief: experimental evidence. PLoS One 10(2):e0118489. https://doi.org/10.1371/journal.pone.0118489

Waldman S (2018) Republican lawmaker: rocks tumbling into ocean causing sea-level rise. In: ScienceMag.org. https://www.sciencemag.org/news/2018/05/republican-lawmaker-rocks-tumbling-ocean-causing-sea-level-rise. Accessed 20 May 20

Chapter 6
Results from South Florida Experiment

Abstract Respondents were randomly assigned either to a control group or to a group that was shown a local map of flooding predicted to occur as a result of a Category 3 hurricane in the respondent's city or county. Based on previous survey research, we had expected the maps to have an impact on perceived risk of sea-level rise, susceptibility of the home to flooding, and support for various mitigation measures. Instead, we found that those exposed to the maps were *less* likely to say they believed that climate change is occurring or that climate change is responsible for more extreme storms. This pattern was especially marked among Republicans. The most important predictors of sea-level rise attitudes were political party identification, gender and education as well as scores on the science questions. For beliefs about impacts on property values, the strongest predictor was political party affiliation.

Keywords Sea-level rise · Flooding and property values · Mitigation

The central purpose of this study was to test the impact of providing dramatic and local visual information about the risk to the individual household and neighborhood on respondent's beliefs about climate change, sea-level rise, and vulnerability to damage and property loss. We randomly assigned respondents either to a control group or to a group that was shown a local map of flooding predicted to occur as a result of a Category 3 hurricane in the respondent's county. We tested the effect of viewing the map on several key outcomes measured: (1) belief that climate change is happening (1 = *definitely not happening* to 7 = *definitely happening*); (2) extent to which climate change is perceived as causing more extreme storms and hurricanes (1 = *not at all* to 5 = *a great deal*); (3) extent to which sea-level rise is seen as connected to climate change (1 = *definitely not related* to 7 = *definitely related*); (4) belief that the sea level is rising (1 = *definitely not rising* to 7 = *definitely is rising*); (5) belief that sea-level rise will affect the respondent's own home (1 = *not at all* to 7 = *a great deal*); (6) belief that sea-level rise will reduce the value of property in the area (1 = *not at all* to 5 = *a great deal*); and, (7) belief that sea-level rise will reduce the value of the respondent's own home (1 = *not at all* to 5 = *a great deal*).

R. Palm, T. Bolsen, *Climate Change and Sea Level Rise in South Florida*, Coastal Research Library 34, https://doi.org/10.1007/978-3-030-32602-9_6

Table 6.1 Effects of map on beliefs about climate change and sea-level rise

	Control (Mean)	Treatment (Mean)	Difference	p-value
Climate change is happening	5.86	5.63	−0.23	0.02
Climate change is causing more intense storms	3.68	3.47	−0.21	0.02
Sea-level rise is related to climate change	5.56	5.42	−0.14	0.17
Sea-level rise is happening	5.50	5.42	−0.09	0.36
Sea-level will affect their own house	4.06	3.83	−0.23	0.05
Sea-level rise will reduce the value of houses in the city	3.75	3.73	−0.02	0.76
Sea-level rise will reduce the value of their own house	3.00	2.77	−0.23	0.07

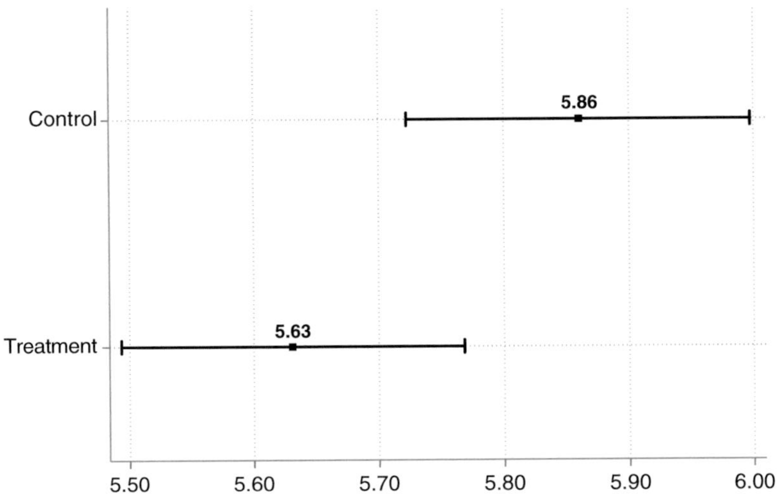

Fig. 6.1 Effect of map treatment on belief in the existence of climate change

Table 6.1 reports the mean difference for each key outcome measure and formal tests for statistical significance in comparing the control and treatment groups. To simplify the presentation of our findings, we employ Figures illustrating the magnitude of the treatment effects on the main outcome measures. We also include, in Appendix 2, a series of ordinary-least-squares (OLS) regressions that estimate the effect of experimental condition assignment on all outcome measures while controlling for other demographic and personal characteristics of the respondents.

The experiment first measured whether or not exposure to the localized flood map would affect respondents' belief that climate change is happening. In contrast to what we predicted, those who viewed a local map of flooding that would result from a Category 3 hurricane had significantly *reduced* perceptions that climate change is happening by nearly one-quarter point on the seven-point response scale we used to measure this belief (Fig. 6.1).

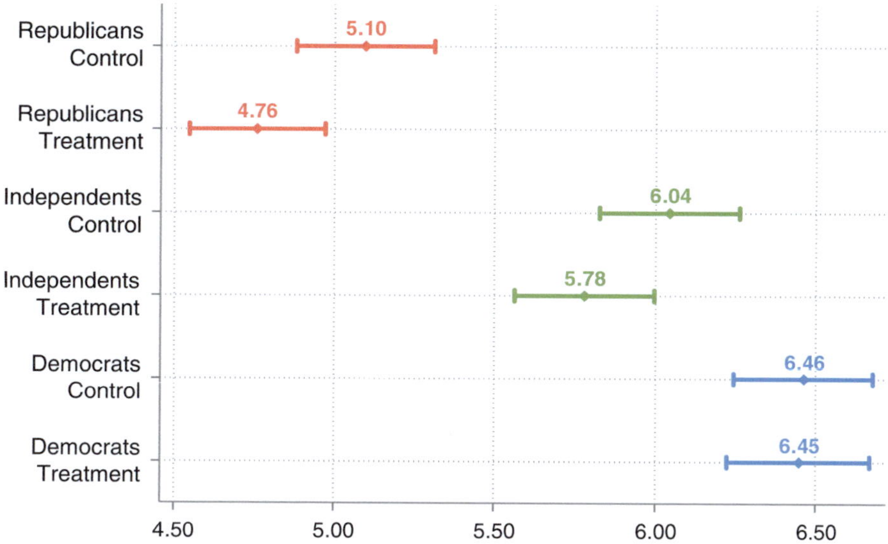

Fig. 6.2 Belief in the existence of climate change

Republicans, and to a smaller degree Independents, drove the negative effect. The mean Republican score was 5.10 in the control condition and the mean score for Independents was 6.04 in the control condition. With exposure to the map, the score decreased to 4.75 for Republicans and 5.78 for Independents (Fig. 6.2). For Democrats, the scores for the two groups were nearly identical.

Similarly, when asked on a five-point scale whether respondents believed that climate change was responsible for more extreme storms and hurricanes, the impact of the map was counterintuitive: those who were exposed to the map were *less* likely to believe that climate change is causing more extreme hurricanes (Fig. 6.3).

Again, exposure to the map elicited the strongest response from Republicans: those exposed to the map scored nearly one-half point lower than those in the control group (Fig. 6.4). The map had little or no impact on Independents or Democrats.

The survey also measured the extent to which coastal residents believe sea-level rise is occurring and the extent to which they perceive sea-level rise as related to climate change. The map had no significant effect *within* partisan categories, but instead differences *between* the partisan groups was striking (Figs. 6.5 and 6.6). The vast majority of Democrats believe that the sea-level is rising and that this is related to climate change but Republicans express *much* greater skepticism about both whether the sea-level is rising and any relationship between sea-level rise and climate change. The majority of Independents believe that sea-level rise is occurring as a result of climate change but are overall less certain than Democrats.

Even if a local map did not impact general beliefs about sea-level rise and climate change, we certainly expected that the map would increase concern about susceptibility of their own home and their own neighborhood to the effects of sea-level rise, as well as potential impacts of this risk on property values. But again, the

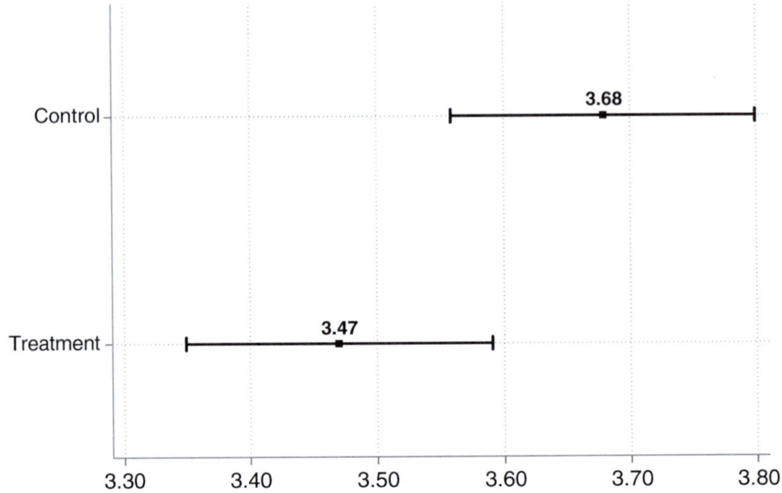

Fig. 6.3 Effect of map treatment on belief climate change causing severe storms

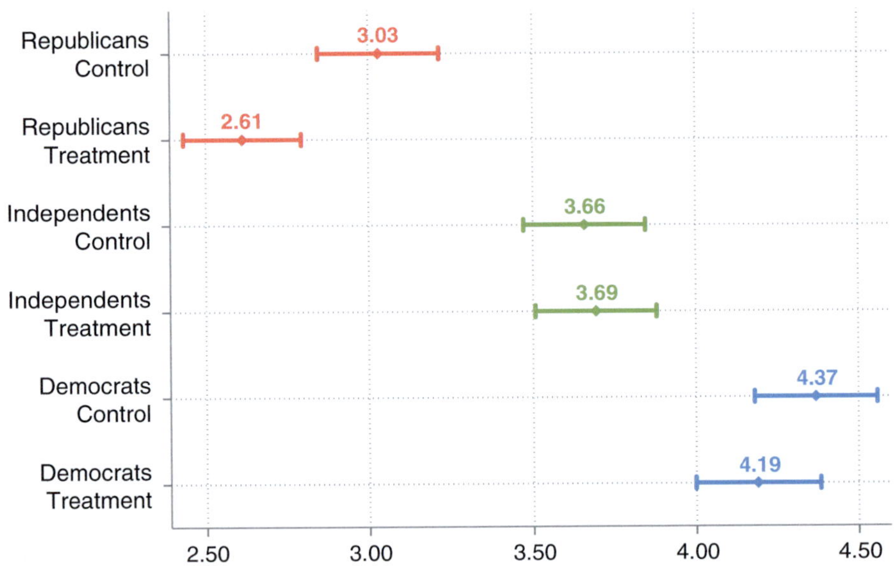

Fig. 6.4 Belief climate change linked to extreme storms

results were counterintuitive. Those who viewed the flood map were less likely to believe that their own home is susceptible to sea-level rise, with a mean score of 4.1 among homeowners in the control condition that received no map compared to 3.7 among homeowners who received the map showing projected flooding from a Category 3 hurricane (Fig. 6.7). Interestingly, we observed this effect for all of the respondents, regardless of political affiliation.

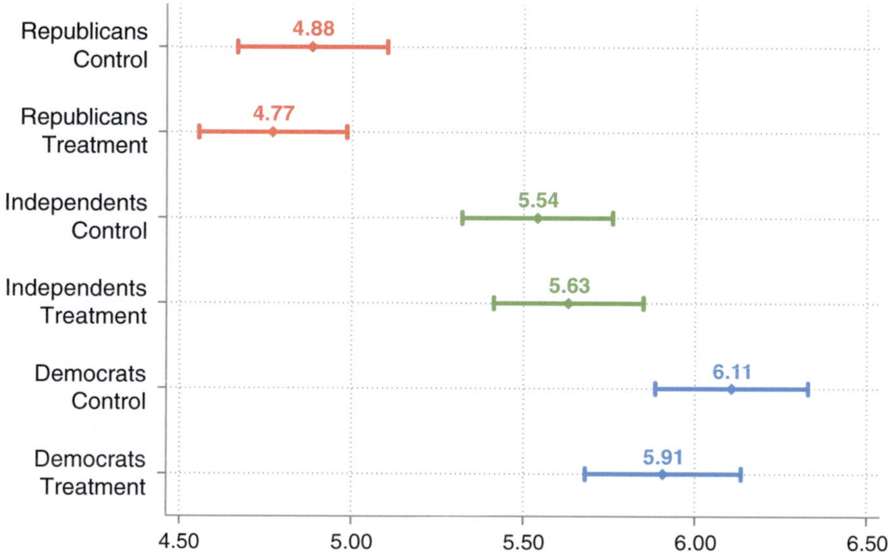

Fig. 6.5 Belief that sea-level is rising

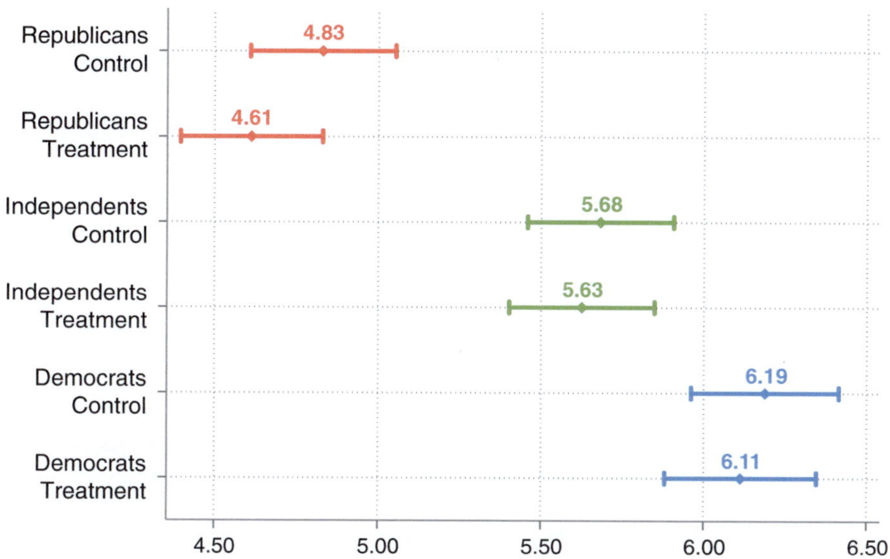

Fig. 6.6 Belief sea-level rise is related to climate change

The map also had no effect on respondents' perceptions about the degree to which sea-level rise will affect property values in respondents' communities. For this variable, the differences between Republicans, Independents, or Democrats are *much* smaller than for the more general beliefs about the existence of climate change

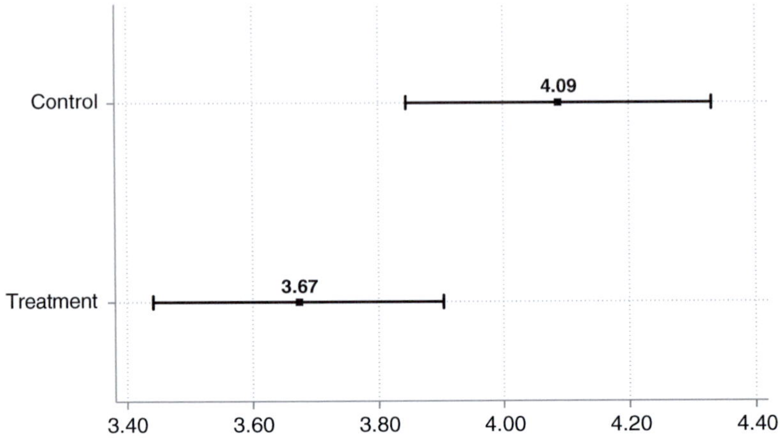

Fig. 6.7 Effect of map treatment on belief home is susceptible to flooding. (Note: Model restricted to homeowners in the sample)

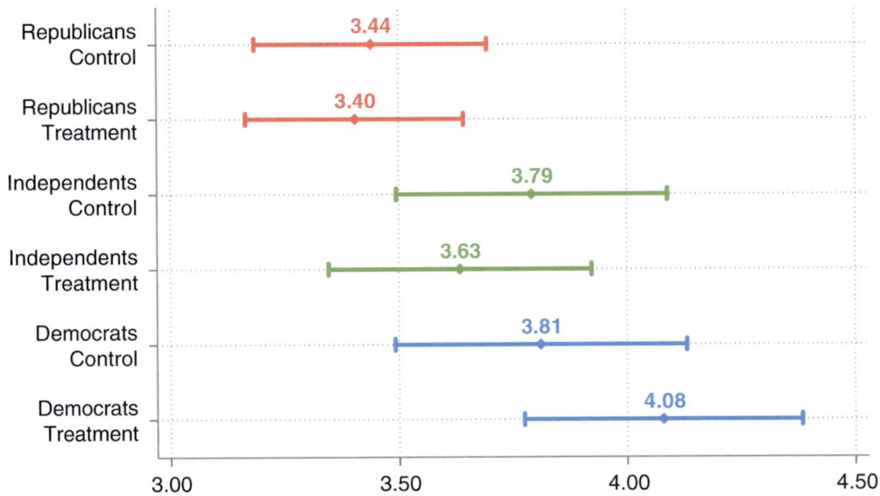

Fig. 6.8 Effect of map treatment on belief sea-level rise will reduce local property values. (Note: Model restricted to homeowners in the sample)

and its relationship to more extreme storms and sea-level rise (Fig. 6.8). To explain this partisan convergence, we posit that material self-interest may be a more powerful motivating force than partisanship.

Finally, and most directly related to homeowner financial interest, we examined how exposure to the flood map would impact homeowners' perceptions that sea-level rise would reduce the value of their own home. Again, counterintuitively, we found that the group who had seen the map had significantly lower scores than the control group with respect to belief that sea-level rise will reduce the value of the

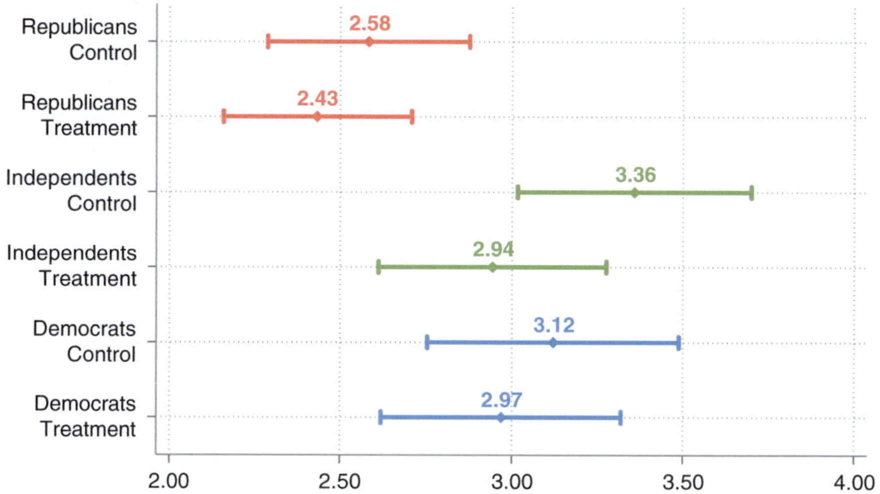

Fig. 6.9 Exposure to map treatment on belief sea-level rise will reduce home value. (Note: Model restricted to homeowners in the sample)

respondents' own home. Regardless of political party identity, all of the respondents expressed uncertainty and some level of skepticism about the impact of sea-level rise on their home's value (Fig. 6.9). The mean score on this outcome measure for Republicans in both the treatment and control condition is well below the mid-point on the response scale, indicating the majority of these homeowners "disagree" with this notion. However, Democratic and Independent homeowners also largely disagree that sea-level rise will reduce the value of their own home, with mean responses for all groups below the scale's midpoint labeled "not sure."

How do we explain these counter-intuitive results? In very different experiments, previous research by Dixon et al. (2019), Hart and Nisbet (2012) and Mildenberger et al. (2019) had shown similar results. In all of these experiments as well as in the one reported here, we believe that the *motivation* of the respondents was a key factor. Motivation affects the way that people process information, especially in the case of a longstanding political issue where there is widespread partisan disagreement. When partisans are exposed to new information that contradicts their existing beliefs, they may process any new information with an underlying goal or motivation to maintain their initial view or uphold their partisan group identity. This can lead them to generate counterarguments against belief-contradictory information, rely on "cues" from trusted sources, and "selectively" search for new information that bolsters their prior view (Kunda 1990; Druckman 2012; Kahan 2015). Alternatively, even when partisans are motivated to form an "accurate" belief—that is, when the underlying motivation or goal is to form a "correct" belief rather than protect existing beliefs, identities or worldviews—they can arrive at very different views in the face of identical information about climate change as a result of differences in the degree to which they perceive the information as credible or trustworthy (Druckman and McGrath 2019). Consequently, people who receive targeted messages about climate change may process and interpret the information in a way that leads to *increased* levels of belief polarization rather than greater consensus.

We conclude from this analysis that the flood map produced either no impact on the way that respondents understand climate change and sea-level rise, or that the effect was counterintuitive: exposure to the map may actually have decreased belief in climate change, decreased belief that climate change is linked to more extreme storms and hurricanes, reduced homeowners' belief that their home is susceptible to sea-level rise or that sea-level rise presents a threat to their property's value. What is perhaps most interesting is that potentially different underlying motivations appear to be at play among coastal residents in our sample depending on the specific climate change outcome measures on which we focus.

Coastal residents' beliefs about the degree to which climate change is happening, whether it is causing more extreme storms and hurricanes, and if sea-level rise is due to climate change are powerfully influenced by the party identification of the respondent. These widespread partisan differences are evident across these beliefs. They largely drive the "unintended" backfire-effect that we observed among Republican respondents who were provided with the information showing local flooding that would result from an extreme weather event and then asked about climate change and its relationship to sea-level rise. On the other hand, in looking at beliefs among homeowners about the degree to which sea-level rise presents a threat to neighborhood and personal property values, the respondents' party identification plays a much smaller role. Democrats, Independents, and Republicans alike express considerable skepticism about whether sea-level rise will have a negative effect on their own property values.

6.1 Effect of Demographic and Political Characteristics on Beliefs and Policy Support

Party identification seemed to override information exposure for many of our key outcome measures, but other demographic and political characteristics are important as well. In the next section, we evaluate the impact of demographic and political characteristics on views about sea-level rise and climate change, the impact of sea-level rise on local property values, and support for policy action to mitigate climate change and the effects of sea-level rise in South Florida. Technical details of this analysis are available in Appendix 2. To simplify the presentation of the results here, we created three scaled measures designed to tap: (1) beliefs about climate change and sea-level rise (combining responses from four individual survey items including belief that climate change is happening, belief that sea-level rise is happening, belief that climate change is causing the sea level to rise, and belief climate change is causing more extreme storms and hurricanes ($\alpha = .91$); (2) beliefs about the effect of sea-level rise on neighborhood property values and for the respondents' own homes (combining responses to questions about homeowners' beliefs about the effects of sea-level rise on local property values and on their own home ($\alpha = .75$); and, (3) support for policy action to address climate change and the effects of sea-level rise in South Florida (combining responses to five survey items dealing with support for specific local polices to address the threat of sea-level rise and general support for climate mitigation ($\alpha = .72$)).

We regressed each of the three scaled measures on the following independent variables: age, income, education, science knowledge, sex, ethnicity (non-white), party identification (Independents are the omitted baseline), ideology (moderates are the omitted baseline), and a "values tradeoff" measure that identifies the degree to which respondents prioritize protecting the environment over economic growth, or vice versa. We also included a dichotomous indicator of whether the respondent was assigned to the experimental treatment condition.

Figure 6.10 plots the coefficient and confidence intervals estimated for the independent variables explaining coastal residents' beliefs about sea-level rise and climate change. First, perhaps unsurprisingly, party identification and political ideology are the most important predictors of residents' views about climate change and sea-level rise. Second, residents who value economic growth over environmental protection are less concerned about climate change and sea-level rise, and less likely to see the two issues as interconnected. Female residents of South Florida were more concerned about climate change and sea-level rise compared to males, a finding that has been replicated in numerous national surveys in the U.S. Finally,

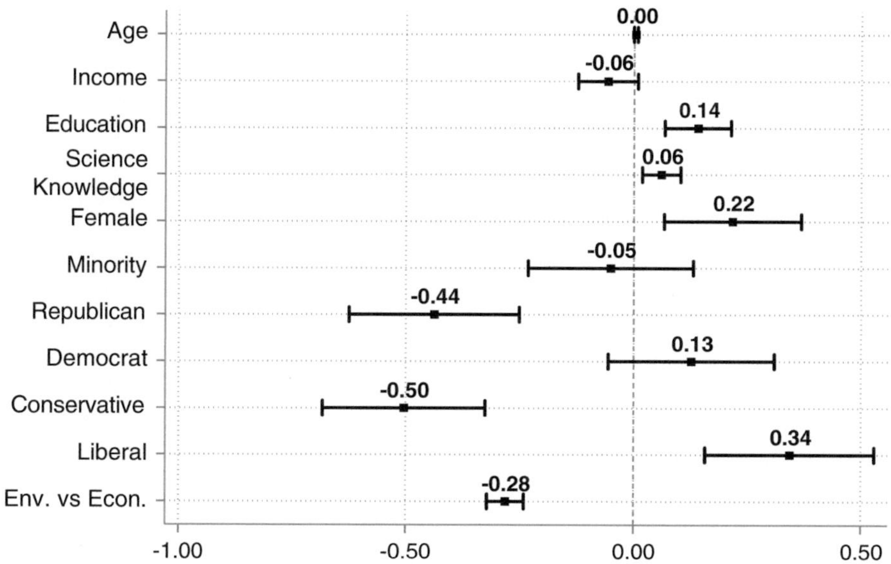

Fig. 6.10 Effects of demographic and political characteristics on beliefs about climate change and sea-level rise

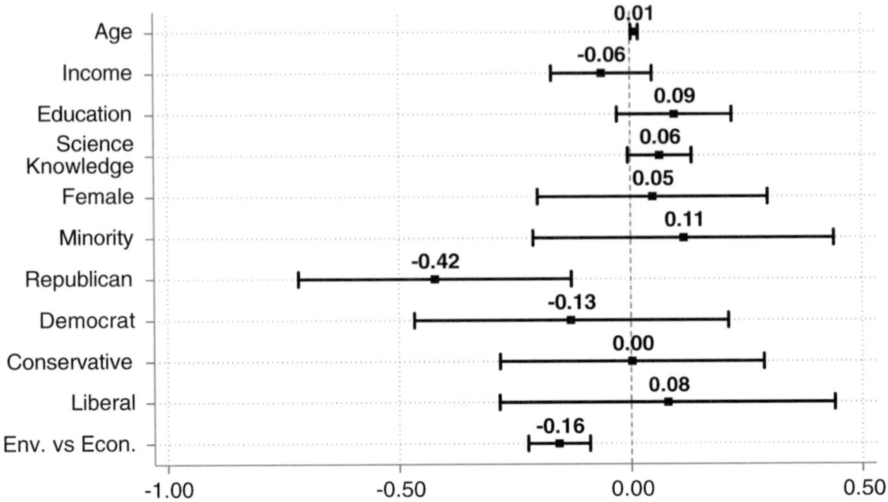

Fig. 6.11 Effects of demographic and political characteristics on beliefs about property values. (Note: Model restricted to homeowners in the sample)

respondents with higher levels of education and who answered more of the science questions correctly on the survey were more likely to believe climate change and sea-level rise are happening.

Figure 6.11 plots the coefficient and confidence intervals estimated for the independent variables in the regression explaining coastal homeowners' scaled beliefs about the effects of sea-level rise on local property values and their homes. Homeowners' beliefs about property values are *not* associated with the respondents' political ideology: although identifying as a Republican is associated with less concern about the impact of sea-level rise on local property values, the impact of partisanship is far less pronounced than when compared to general beliefs about climate change and sea-level rise. Values associated with the protection of the environment as opposed to maintaining economic growth also are strongly associated with beliefs about the degree to which sea-level rise will have a negative effect on local property values. Other demographic factors with the exception of education and science awareness appear to be largely unrelated to these beliefs: those who report more levels of school completed and those who possess higher levels of objective knowledge about science also report slightly greater concern about the effect of sea-level rise on property values.

The factors that are associated with coastal residents' support for legislative action to mitigate the effects of sea-level rise and climate change are powerfully influenced by party identification and political ideology (Fig. 6.12). Republicans and conservatives express *far less* support for such policies, while females are relatively more supportive of polices to address the threats posed by sea-level rise and climate change. Residents with higher levels of formal education are also more likely to express support for policy action, but higher levels of science knowledge is

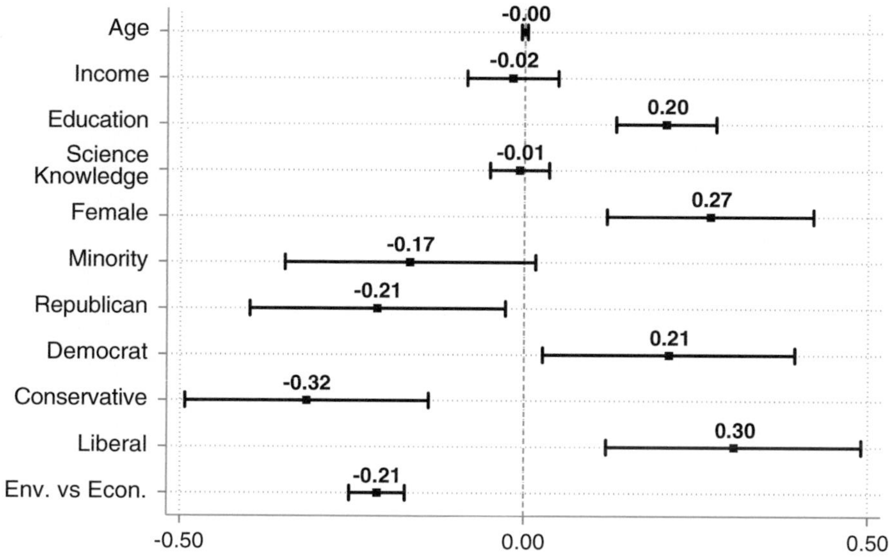

Fig. 6.12 Effects of demographic and political characteristics on support for policies to mitigate the effects of climate change and sea-level rise

unrelated. Favoring protection of the environment over economic growth is also associated with support for mitigation policy.

6.2 Summary

The experimental part of the survey presented maps to half of the respondents. These maps portrayed the flooding that would be associated with a Category 3 Hurricane 15 years from now. Each respondent in the treated group was shown a map of their own city or county, usually quite close to their own home. We had expected the maps to have an impact on perceived risk of sea-level rise, the susceptibility of the home to flooding and even support for various mitigation measures.

What we found instead was that those exposed to the maps were less likely to say they believed that climate change is occurring or that climate change is responsible for more extreme storms. This pattern was especially marked among Republicans. The maps had no impact on belief about sea-level rise. When we asked owner-occupiers whether their homes were susceptible to sea-level rise or whether the property values in their city or their home would be affected by sea-level rise, those exposed to the maps were less likely to state they were at risk. For these questions, involving personal risk to the respondents and their home values, there was a weaker partisan differential.

Although the map experiment did not increase awareness or concern with sea-level rise, we found several consistent correlates of risk perception. For sea-level rise, the most important predictors were political party identification, gender (female) and higher education levels, as well as higher scores on the science questions. For property values, the strongest predictor was political party affiliation, although this effect was weaker than for other climate change impacts. For general support of mitigation legislation, party, political ideology, gender and education level were the most important predictors.

The statistical analysis just presented gives an overview of the factors associated with the acceptance of sea-level rise affecting South Florida and beliefs of the individual residents overall. In the next section, we will review in some detail five case studies of respondents who were shown maps of likely flooding with a hurricane Category 3 and are nonetheless certain that their own home is not at risk.

References

Dixon G, Bullock O, Adams D (2019) Unintended effects of emphasizing the role of climate change in recent natural disasters. Environ Commun. Advance online publication. https://doi.org/10.1080/17524032.2018.1546202

Druckman JN (2012) The politics of motivation. Crit Rev 24(2):199–216. https://doi.org/10.1080/08913811.2012.711022

Druckman JN, McGrath MC (2019) The evidence for motivated reasoning in climate change preference formation. Nat Clim Change 9:111–119. https://doi.org/10.1038/s41558-018-0360-1

Hart PS, Nisbet EC (2012) Boomerang effects in science communication: How motivated reasoning and identity cues amplify opinion polarization about climate mitigation policies. Communic Res 39(6):701–723. https://doi.org/10.1177/0093650211416646

Kahan DM (2015) Climate-science communication and the measurement problem. Polit Psychol 36:1–43. https://doi.org/10.1111/pops.12244

Kunda Z (1990) The case for motivated reasoning. Psychol Bull 108(3):480–498. https://doi.org/10.1037/0033-2909.108.3.480

Mildenberger M, Lubell M, Hummel M (2019) Personalized risk messaging can reduce climate concerns. Global Environ Chang 55:15–24. https://doi.org/10.1016/j.gloenvcha.2019.01.002

Chapter 7
Denial of Risk

Abstract Five examples of owner-occupiers of single-family detached dwellings illustrate the variability in age, party identification, ideology and education of some respondents whose homes actually are at risk but who view their property as not susceptible to sea-level rise. What becomes clear from the description of these respondents is that the partisan filter through which South Florida residents view sea-level rise and their own vulnerability is overwhelming.

Keywords Case study · Sea-level rise · Flood susceptibility

Individuals reach decisions and form opinions through very complex processes that are difficult to capture through an analysis of percentages or statistics. Each is a unique combination of demographic and economic characteristics, attitudes and life histories. In order to get a better sense of the people behind our statistics, we selected five cases for a more detailed description. We have anonymized these cases, presenting only general information about where the respondents live and who they are. Each of the respondents was an owner-occupier of a single-family dwelling. Each was presented with a map of their own city or county showing inundation associated with a Category 3 hurricane. In each case, examination of their particular location using FloodIQ.com clearly indicates that their property or that of their neighbors is at serious risk of flooding in the event of such a hurricane, and in some cases is already susceptible to flooding simply from high tides. Yet, in each case, the respondent reported that his or her home is "definitely not" susceptible to sea-level rise.

The youngest of the five is a Latina female who is about 35 years old. She has a college degree and her earnings are in the middle-income range of this study ($30,000–$69,900). She purchased her home just 2 years ago. Homes in her neighborhood list at around $275,000–$350,000, were built in the 1950s and 1960s, and have about 111.48–139.35 square meters (1200–1500 square feet) of living space. She is the only one of the five cases who has a separate flood insurance policy on her home. She considers herself to be "somewhat conservative" and leans Republican. When considering trade-offs between protecting the environment or protecting the economy, she would "very likely protect the environment." She

R. Palm, T. Bolsen, *Climate Change and Sea Level Rise in South Florida*,
Coastal Research Library 34, https://doi.org/10.1007/978-3-030-32602-9_7

slightly agrees with the assertions that "we have gone too far in pushing equal rights in this country" and also that "the government should spend less time trying to solve people's problems." She trusts science to solve problems "to a moderate extent," and correctly answered all six of the science knowledge questions. She believes that sea-level rise is "very likely" taking place, and that sea-level rise would reduce property values in her city and her own house a moderate amount. She believes that climate change is "very likely happening," and that it is moderately related to an increase in extreme storms and hurricanes but is "not sure" whether sea-level rise is related to climate change. She supports each of the five mitigation measures suggested in the questionnaire. She says she is "extremely concerned" about the effects of climate change and feels "a great deal" of anxiety. She slightly agrees that her "own actions can affect climate change" and strongly agrees that it is "the responsibility of the government rather than citizens" to deal with this issue. She is "very interested" in learning more about sea-level rise, but moderately agrees that research on climate change is politically motivated. She also moderately agrees that scientists think climate change is occurring because of human action and is neutral about whether or not the entire issue of human-induced climate change is a hoax.

She was shown a map of Fort Lauderdale (Fig. 7.1). In comparison, Fig. 7.2 shows the neighborhood of this respondent, and the highest annual tidal flood level for this year, even without further sea-level rise. In a Category 3 hurricane, her house and neighborhood would be completely inundated, yet she maintains that her home is "definitely not" at risk.

Fig. 7.1 Questionnaire map of Fort Lauderdale. (Source: FloodIQ.com)

Fig. 7.2 High tide neighborhood of respondent 1. (Source: FloodIQ.com)

The second respondent is a 60-plus year-old white woman from the town of Dunedin, just north of Clearwater on the Gulf coast. She has an above average income as defined for the purposes of this study ($70,000–$99,999) and attended some college but did not earn a degree. She just recently purchased her home, although she has lived in Florida for more than 30 years. Homes in her neighborhood list at around $280,000–$380,000; many were built in the 1970s and have about 130.06–195.096 square meters (1400–2100 square feet) of living space. She says she is "somewhat liberal" and leans Democrat. She considers protecting the environment and promoting the economy as of equal importance, and she strongly disagrees with the assertion that "we have gone too far in pushing equal rights in this country." She is neutral on the issue of whether or not "the government should spend less time trying to solve people's problems." She did moderately well on the science questions, answering four of the six correctly, and trusts in science "to a very great extent" to overcome problems. She believes that sea-level rise is "probably" taking place, and that sea-level rise would reduce property values in her city and her own house "a moderate amount." She believes that climate change is "definitely happening," and also strongly believes that sea-level rise is related to climate change. She feels that although the property values for Dunedin will be reduced "a great deal" if sea-level rise occurs, it would not affect the value of her own house very much. She enthusiastically supports mitigation measures such as zoning regulations and the construction of sea walls but is less enthusiastic about taxing carbon-emitting companies and is "strongly opposed" to increasing the price of fossil fuels. She is "somewhat concerned" about the effects of climate change but feels only "a little" anxiety about sea-level rise. She does not believe that her actions have any influence on climate. She is "very interested" in learning more about sea-level rise, and "moderately disagrees" with the statement that research on climate change is politically motivated. She "strongly agrees" that scientists think climate change is occurring because of human action. She was shown a map of the St. Petersburg area as it would be affected by a Category 3 hurricane in 2033 (Fig. 7.3).

Although she indicated that her home was "definitely not at risk," Fig. 7.4 illustrates the projected damage to her neighborhood from a Category 3 hurricane in her area in 15 years: all of coastal Dunedin would be inundated as well as the connection to the neighboring barrier islands (Fig. 7.4).

The third sample respondent is an Asian-American woman in her early 50s who lives in Riverview, a suburb of Tampa on the Gulf Coast. Her income is in the highest range of the five case studies ($100,000–$200,000) and she also has the highest level of education completed (an advanced degree). She purchased her home in Riverview about 5 years ago, although she has lived in Florida for about 30 years. Single family detached homes in her neighborhood list at around $200,000–$275,000, and many have been built quite recently. She considers herself to be a strong Republican although "somewhat liberal." When considering trade-offs

Fig. 7.3 Questionnaire map of St. Petersburg, Florida. (Source: FloodIQ.com)

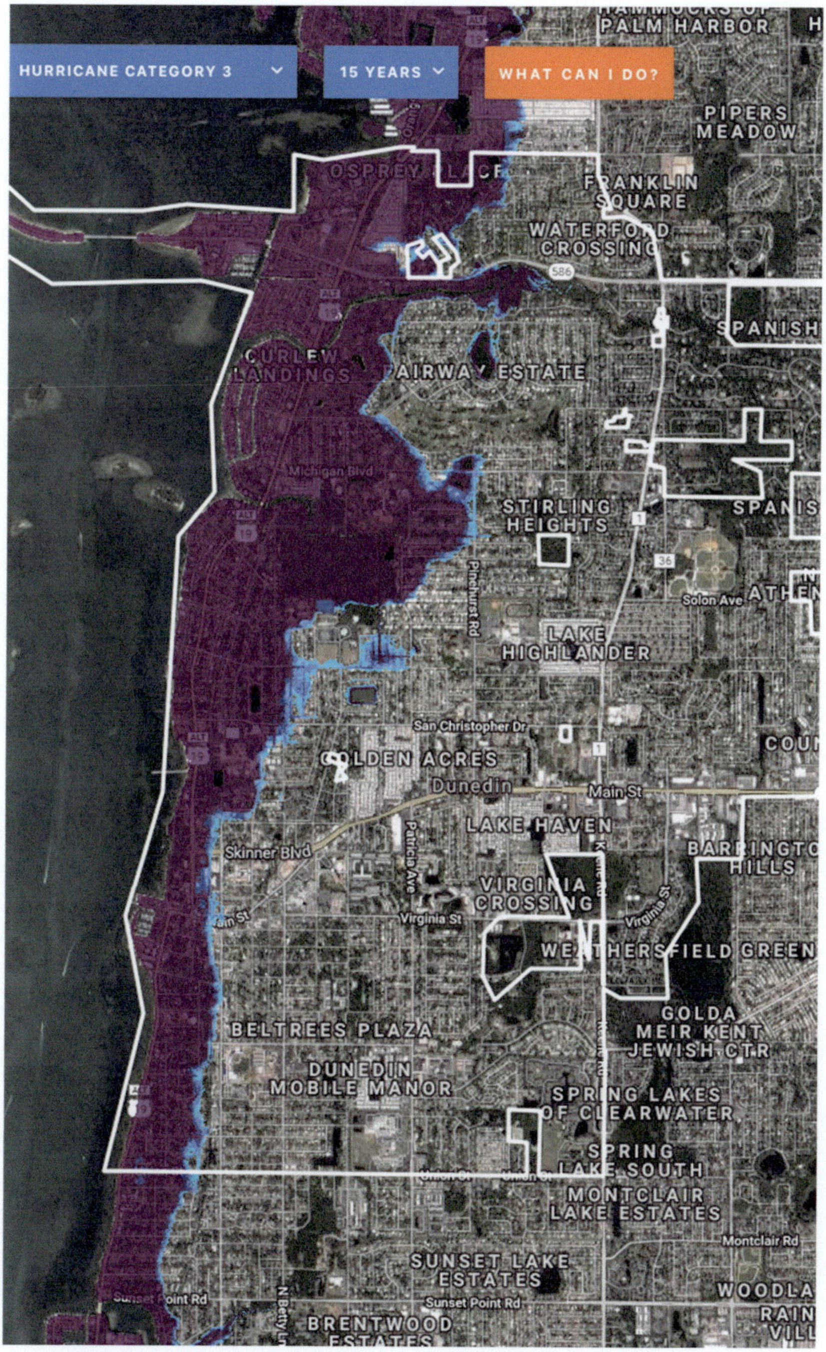

Fig. 7.4 Map of flooding in neighborhood of respondent 2. (Source: FloodIQ.com)

between "protecting the environment or protecting the economy," she would "definitely protect the environment," a perspective often not linked with conservatives. She is neutral with respect the assertion that we have "gone too far in pushing equal rights in this country" and strongly agrees that "the government should spend less time trying to solve people's problems." She trusts science to solve problems "to a great extent," and correctly answered all six of the science knowledge questions. She believes that sea-level rise is "probably happening," and that sea-level rise would reduce property values in her city and her own home "a great deal." She believes that climate change is "definitely happening," and that sea-level rise is related to climate change. She also believes that climate change is causing more extreme storms and hurricanes. She supports zoning laws to increase set-backs from the coast, the construction of sea walls or barriers, and the taxation of companies that produce greenhouse gases. She "strongly opposes," however, programs in which the government buys back land from people who live in flooded areas and "strongly opposes" an increase of price on fossil fuels to encourage energy conservation. She says she is "very concerned" about the effects of climate change, but only feels "a little anxiety" about sea-level rise. She "moderately disagrees" that her own actions can affect climate change and "strongly disagrees" that "it is the responsibility of the government rather than citizens" to deal with this issue. She is "very interested" in learning more about sea-level rise, and "strongly disagrees" that research on climate change is politically motivated. She "strongly agrees" that scientists think climate change is occurring because of human action, and "strongly disagrees" that the entire issue of human-caused climate change is a hoax. The map that she saw (Fig. 7.5), focused on downtown Tampa.

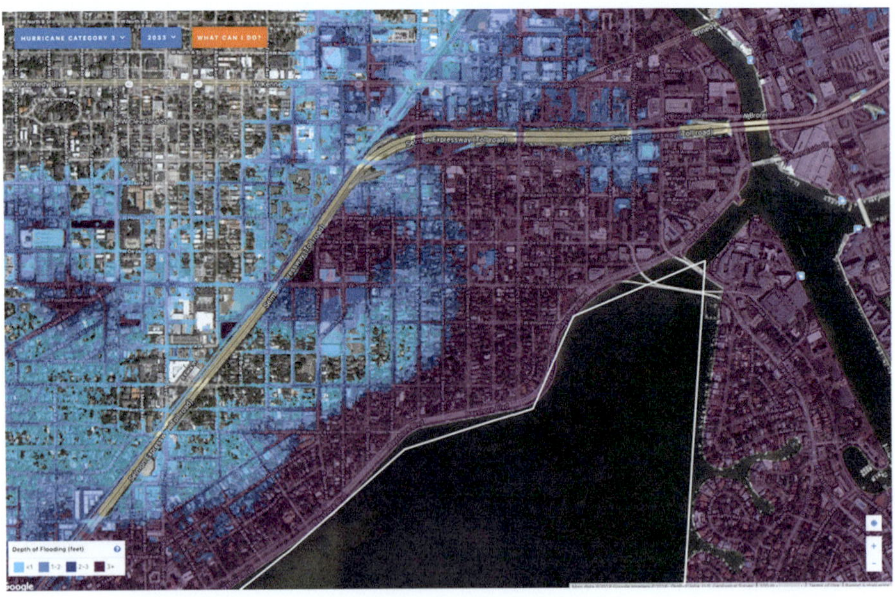

Fig. 7.5 Questionnaire map of Tampa. (Source: FloodIQ.com)

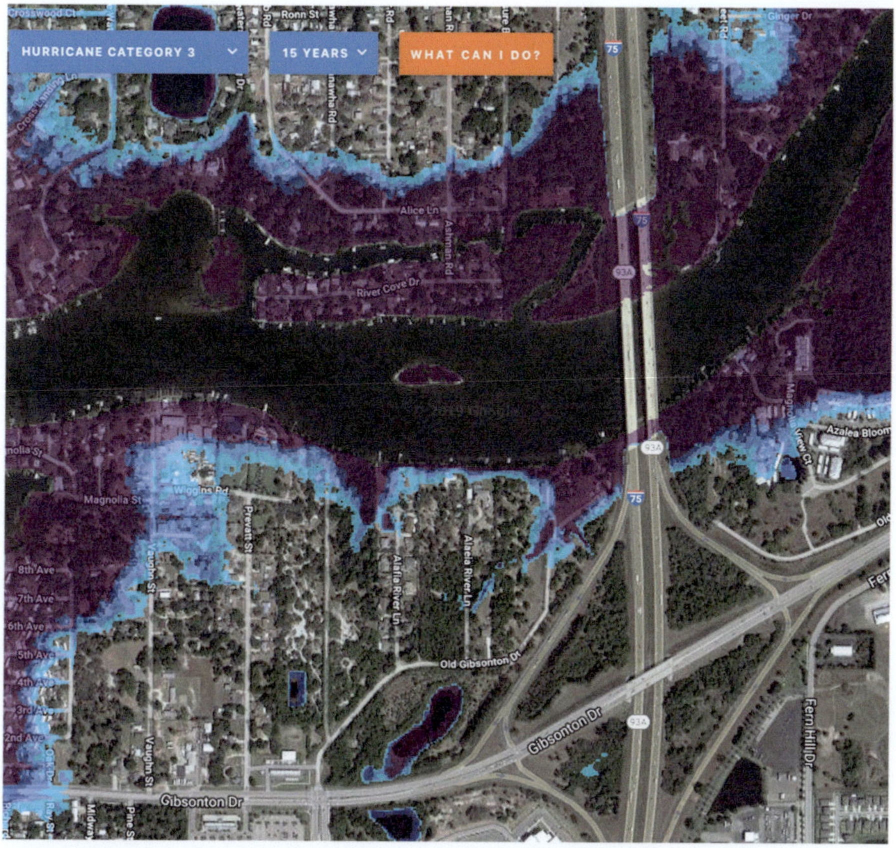

Fig. 7.6 Flooding near the home of respondent 3. (Source: FloodIQ.com)

The map below (Fig. 7.6) portrays the location of her neighborhood with respect to flooding in a category 3 hurricane. Clearly the neighborhood would suffer severe damage with dramatic levels of flooding.

The fourth respondent who also lives in an area susceptible to flooding and who also denies the vulnerability of his home to sea-level rise was a white man in his early 70s. He lives in Cape Coral, a city that was described earlier as having been built to create a series of waterfront residential lots on canals dredged from the previous wetlands. He has an above-average income ($70,000–$99,999), has completed some college, and has lived in his home for over 15 years. Single family detached homes in this neighborhood list for anything from $200,000 to $650,000 depending on the amenities of the house and neighborhood. Many have been built recently. He is a "strong Republican" and "very conservative." When considering trade-offs between protecting the environment or protecting the economy, he views the two issues as equally important. He "strongly agrees" both that we have "gone too far in pushing equal rights in this country" and that "the government should spend less time trying to solve people's problems." He trusts science only to a small extent but did well in answering the science questions, having gotten five of the six answers correct. He believes that the sea-level is "very likely" *not* rising, that climate change is "definitely" not happening, and that if there were sea-level rise, it would have no impact on either the value of his own house or on property values in Cape Coral. He supports none of the mitigation policies presented in the questionnaire, is unconcerned with climate change, has no anxiety about sea-level rise, is not interested in learning more about sea-level rise and disagrees that his own actions can affect climate change. He "strongly disagrees" that most scientists think that climate change is the result of human activities and instead "strongly agrees" that the idea of climate change as related to human activity is a hoax or conspiracy. The map that he was shown focused on Fort Myers (Fig. 7.7). In a Category 3 hurricane, the neighborhood in which he lives would be under more than three feet of water (Fig. 7.8).

The final example is also a white man in his mid-60s. He is a college graduate and has an average income level. He is "mostly conservative" and "leans Republican." He purchased his home in Tampa nearly 10 years ago but has lived in Florida for more than 30 years. Houses listed for sale in his neighborhood range in price from $200,000 to $350,000. He did well on the science questions, answering five of the six correctly, but has little faith in science to solve problems. He views the environment and the economy as equally important issues, and slightly agrees both that "we have gone too far in pushing equal rights" and that "the government should spend less time trying to fix everyone's problems." He feels that the sea-level "probably is not rising," and that if it does rise, it may reduce property values "a little" in Tampa

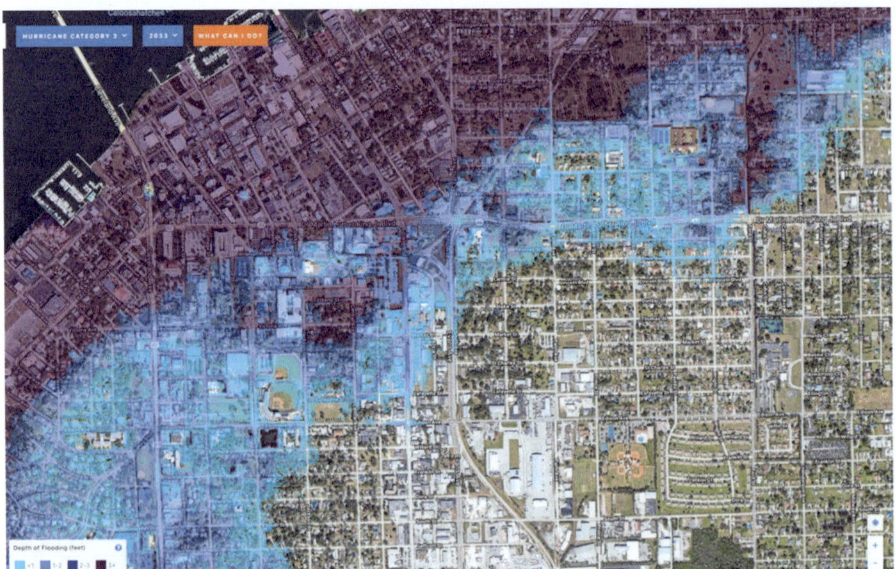

Fig. 7.7 Questionnaire map of Fort Myers, Florida. (Source: FloodIQ.com)

Fig. 7.8 Map of flooding in neighborhood of respondent 4. (Source: FloodIQ.com)

but will not affect the value of his house at all. He believes that climate change is also "probably not happening," but that sea-level rise "probably" is related to climate change, and that climate change has a small influence on causing more extreme storms and hurricanes. He supports somewhat each of the five mitigation measures presented, he is "somewhat concerned" about the effects of climate change but has no anxiety at all about sea-level rise. He "moderately disagrees" that individual actions have an influence on climate change and "slightly agrees" that governments have the primary responsibility for dealing with climate change. He is indifferent

Fig. 7.9 Map of neighborhood of respondent 5 (Source: FloodIQ.com)

about learning more about the effects of sea-level rise, and "moderately agrees" that most scientific research on climate change is politically motivated. He "slightly disagrees" that most scientists think climate change is happening because of human activities and also "slightly disagrees" that the idea of human induced climate change is a hoax or conspiracy. In the course of the survey, he was shown the Tampa map (Fig. 7.5). The map of his neighborhood (Fig. 7.9) indicates that most of the residential property and streets in his neighborhood would experience 30.38–91.44 centimeters (1–3 feet) of flooding in a Category 3 hurricane, clearly belying his confidence that his property is safe.

7.1 Discussion

We designed this study with high expectations that providing homeowners and other residents of south Florida with local full-colored maps showing the likely inundation of land in their communities associated with a Category 3 hurricane in 2033 would have an impact on their concerns with climate change, sea-level rise, and the risk to their own homes and communities. Instead, we found that providing these maps had a counterintuitive effect: those who were exposed to the maps were less likely to show concern for these issues than the control population who did not receive the maps. The fact that the information we presented is also publicly available makes this finding even more remarkable since at most, the impact of prior exposure to such information would have produced a null effect rather than a significant negative effect. The partisan filter through which South Florida residents view sea-level rise and their own vulnerability is overwhelming. The one issue in which such partisan differences is reduced is the threat to home equity: for this issue, despite evidence to the contrary, many respondents expressed *skepticism* that their *own* home value would be affected.

Finally, although there have been many laboratory experiments about providing information to various groups and testing their responses, few studies have conducted this type of survey experiment on a population particularly vulnerable to hazards of this nature. The findings from this experiment point to the fact that it is difficult to predict the kinds of information residents will actually use to assess their own vulnerability.

Chapter 8
Prospects for Mitigation

Abstract Over the past 20 years, Florida's political leadership has promoted distinctly different policy responses with respect to sea-level rise mitigation. In the first decade of the current century, a Governor's Action Team on Energy and Climate Change outlined specific policy recommendations to increase energy efficiency, including the establishment of a "cap-and-trade" program. More recently there has been far less enthusiasm for state government intervention and at times even denial of climate change. The survey research reported here indicates that those residents living in areas susceptible to coastal or riverine flooding are aware of their risks and generally support new regulations requiring greater set-backs from the coast and the construction of flood barriers or sea walls. They are less enthusiastic about measures that would involve increased costs to the individual such as increasing taxes on fossil fuels to encourage energy conservation.

Keywords National Flood Insurance Program · Building codes · Flood barriers · Cap-and-trade

8.1 Legislation in Florida to Mitigate Impacts of Sea-Level Rise and Climate Change

What are some of the measures that have been adopted in Florida? At the state level, responses to climate change were initially linked to energy policies (Ruppert and Deady 2017). Former Governor Charlie Crist, at the time a Republican, issued executive orders to reduce greenhouse gas emissions in 2007, and the legislature created a Florida Energy and Climate Commission in 2009. A report to the governor dated October 15, 2008 from the Governor's Action Team on Energy and Climate Change (Center for Climate Strategies 2008) outlined fifty recommendations to increase energy efficiency, invest in low-carbon energy sources and guide a "low-carbon economy," as well as develop the "cap-and-trade" program that was already in existence in the state. These recommendations were both specific, such as the adoption

R. Palm, T. Bolsen, *Climate Change and Sea Level Rise in South Florida*, Coastal Research Library 34, https://doi.org/10.1007/978-3-030-32602-9_8

of "low rolling resistance tires" as well as more general, such as the admonition to create general campaigns for "public awareness and education."

Governor Crist was succeeded by Rick Scott in 2011, and the emphasis of state government shifted from growth management to economic development. In the area of climate change, state law (FS., Ch. 163) was changed to remove many of the greenhouse gas reduction provisions (Ruppert and Deady 2017). The initiative to create new regulations to mitigate either climate change or sea-level rise shifted responsibility from the state to the local governments. The result is a patchwork of regulations all within the context of a general state mandate that local governments include coastal management in their comprehensive plans.

The change in the federal flood insurance program in 2012, drastically increasing the price of insuring coastal property through the National Flood Insurance Program, had major impacts on Florida. The state had to deal with the new flood insurance regulations as well as issues of disaster recovery. In 2015, Florida's Peril of Flood Act (F.S. 163.3178.2.f.1.) enhanced requirements for using engineering solutions to flooding, tightened construction requirements within floodplains, and encouraged local governments to participate in the National Flood Insurance Program community rating system.

Local governments responded to federal and state requirements in a variety of ways, many incorporating sea-level rise into their comprehensive plans and, in addition, even adding "optional elements" concerning the reduction of greenhouse gases. However, even all of these efforts have been described as "far from sufficient to build a more resilient future for Florida" (Ruppert and Deady 2017). State and local governments have focused on building standards rather than land use (Southeast Florida Regional Compact 2017). They are also using estimates of rainfall, storms and erosion derived from past history rather than projections for accelerated risk associated with future climatic change.

Even more importantly, there has been little follow-up to the well-intentioned measures undertaken during the Crist administration to address the root causes of climate change and sea-level rise. Governor Scott, elected to the U.S. Senate in 2018, is generally not seen as an advocate for an active response to climate change. In an analysis of the "Rick Scott Effect" with respect to enforcement of environmental cases, Lynch et al. (2019) concluded that enforcement of environmental regulations sharply declined during Scott's tenure as governor. Although he has denied this, Governor Scott is even said to have instructed state agencies not to use the terms "climate change" or "global warming" in documents (Allen 2015). The Lynch et al. (2019) analysis revealed that these words had almost disappeared from state reports by 2014. In contrast, Ron DeSantis, elected in 2018, has mentioned the term "climate change" and established a new Florida Office of Resilience and Coastal Protection (Green 2019; Florida Department of Environmental Protection 2019). However, it is still unclear that Florida's political leadership will embrace the kind of mitigation platform that was in effect in the Crist administration, and it is highly unlikely that the current (2019) federal administration will push Florida in that direction.

8.2 General Support for Additional Measures to Mitigate Climate Change and Vulnerability to Sea-Level Rise

The survey of South Florida residents living in postal codes susceptible to coastal or riverine flooding showed that there is general awareness of sea-level rise, although with partisan differences in opinion. Only a minority of the coastal population is not sure or is convinced that it is not rising. Such a majority should be able to influence public policy at least locally and create an environment that would support changes in individual behavior and environmental design that could reduce vulnerability to sea-level rise. But does this awareness of the problem translate into policy support?

We investigated support for five measures that could mitigate damage or slow the process of climate change. The first was changes to zoning laws that would require that new construction is located further inland from the coast to protect it from tidal flooding. We asked about this measure in a straightforward manner, without bringing up issues around specific elevations that would also normally accompany such a change in zoning standards. Florida currently controls construction along ocean beaches through a combination of local zoning and a statewide Coastal Control Construction Line or "CCCL" (Building a Safer Florida Inc. 2017). The 2017 requirements of this part of the Florida Building Code "delineates the inland limit of areas where special siting and design considerations are necessary to protect the beach-dune system and to protect proposed and existing structures, adjacent properties, and the preservation of public beach access" (FBC, Sec. 3109 coastal construction control line requirements. In: 6th ed. Florida Building Code). The law deals with issues such as the elevation of the first floor of habitable space, which must be at least 30.48 centimeters (1 foot) above the base flood elevation (BFE) as defined by FEMA flood insurance maps. Flood insurance premiums are reduced for every centimeter above this level up to 1.219 meters (4 feet). In counties where no CCCL has been established, coastal construction is "prohibited within 50 feet [15.24 meters] of the line of mean high water" (F.S. 161.053). Since home construction and its regulation has long been a part Florida's culture, a change in state or county laws, ordinances, or regulations to increase the distance from the coast where housing can be built is a modification in practice rather than an entirely new idea.

For the Florida respondents, the modal position on requiring greater set-backs from the coast is neither opposition nor support, but overall there is much more support than opposition from advocates of both political parties. As with many other questions we asked, Democrats are more likely to support or strongly support such new regulations than are Republicans (Fig. 8.1)

A second potential mitigation against future flooding would be a program in which the government buys back land from people who live in areas that will flood. The purpose of such a program is to remove flood-prone houses from the market to ensure that the same homes are not resold only to be flooded again. This type of program was adopted in Houston after Hurricane Harvey in 2018 when the federal government through the Federal Emergency Management Agency along with the Texas General Land Office made mitigation grants to Harris County, Texas to per-

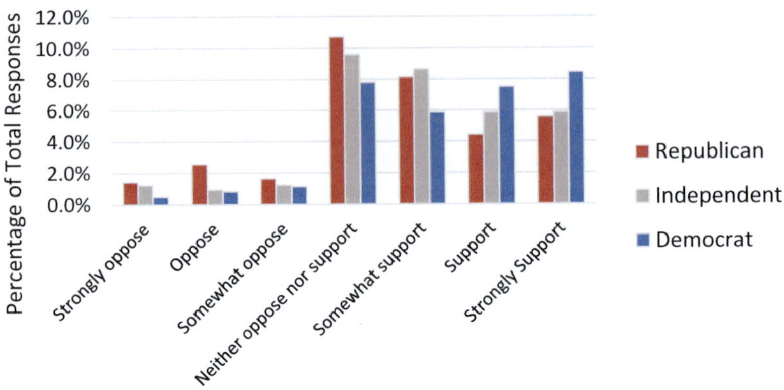

Fig. 8.1 Support for zoning laws requiring greater set-backs

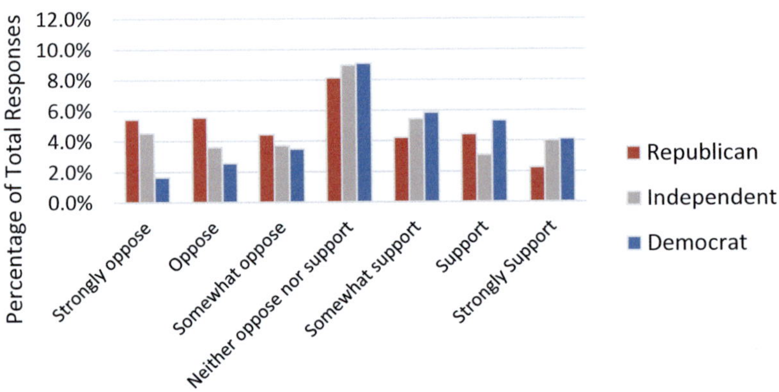

Fig. 8.2 Support for a government program to buy-out property at risk

mit the county to purchase homes that are susceptible to repeated flooding (Cusick 2018). It was a voluntary program in the sense that owners were not required to sell to the government. The South Florida respondents were generally not enthusiastic about this measure, with only 19.4% of Republicans and 29.5% of Democrats expressing either "support" or "strong support" (Fig. 8.2). This type of policy is relatively uncommon and would require a change in thinking about the role of government in indemnifying homeowners at risk.

A third measure we asked about is the commonplace practice of erecting flood barriers to divert flood waters. Such barriers may include the familiar levees along rivers, removable or demountable flood walls that can be put in place when there is warning of a flood, or more permanent system of sea walls, elevated roads and a pumping system. Miami Beach, for example, has invested $500 million to install pumps and raise roads (Flechas 2017). A system of barriers or sea walls has also been constructed in various parts of the state. In some cases, the sea wall construction is undertaken voluntarily by the property owner (Fig. 8.3). In other cases, municipalities or even counties may require such construction (Barszewski 2018).

The construction of sea walls, although an expensive option to build and maintain, seems to be effective, at least for the short term and in the immediate areas they are built to protect. They are sometimes opposed on aesthetic grounds, and more significantly because sea walls displace erosion, causing even greater effects downdraft of the seawall (Balaji et al. 2017; Woolfenden 2013). However, since construction of sea walls is an established practice, it is probably not surprising that it had little opposition from respondents of either party, with 64.2% of the respondents overall expressing support for such a policy (Fig. 8.4).

We also posed two questions on support for measures to deal not with the immediate issue of flooding, but instead the more general issue of climate change. The first was whether the federal government should tax companies based on the amount of greenhouse gases they emit. This policy is commonly known as a "carbon tax" and has been proposed in Congress several times. According to the Environmental Protection Agency (EPA), more than three-quarters of greenhouse gas emissions are caused by transportation, the generation of electricity and industry (EPA 2017). In an analysis by the Congressional Budget Office (2018), the imposition of a tax of $25 per metric ton on emissions of greenhouse gases from electricity generation, manufacturing and transportation, would reduce U.S. emissions of greenhouse gases in a cost-effective way, providing incentives for business and households to reduce emissions while increasing federal revenues by over a billion dollars (Congressional Budget Office 2018).

Fig. 8.3 Sea wall construction. (Source: USGS)

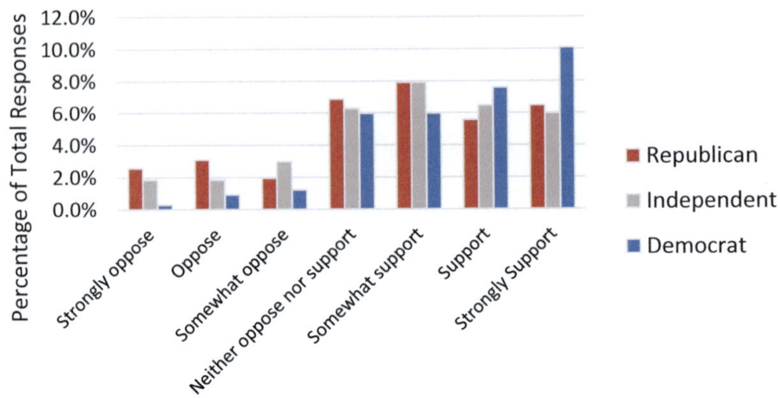

Fig. 8.4 Support for government construction of flood barriers

Florida respondents generally favored the carbon tax, but with very large differences in support by political party. Overall, 63% of the respondents slightly, moderately, or strongly support a carbon tax. The corresponding percentages were 46.2 for Republicans and 78.8% for Democrats. A survey by the National Opinion Research Center taken at about the same time as the Florida survey and phrased similarly ("would you support, oppose, or neither support nor oppose a policy to reduce greenhouse gas emissions by taxing the use of carbon based fuels such as coal, oil, and natural gas based on how much they contribute to climate change") showed about 44% of this sample in support of a carbon tax (Energy Policy Institute 2018). A similar level of support can be deduced from the voting patterns in the state of Washington where a referendum to create the first carbon tax in the nation was backed by only 44% of voters (Gleckman 2018). A simple comparison of these numbers suggests that overall, the Florida coastal residents seem more supportive of a carbon tax than the general US population.

The second climate change mitigation question asked about their own personal commitment to reducing carbon consumption by paying higher taxes on fossil fuels such as gasoline. We asked whether the price of these fuels should be increased in order to encourage saving energy. Responses showed extreme partisan division of opinion (Fig. 8.5). Overall 39.3% of the respondents at least slightly supported such a policy, but 54.2% of the Democrats and only 24.9% of the Republicans were supportive.

8.3 Some Final Words

The survey reported in this monograph took place over the period of October 24 to December 17 of 2018. During the weeks immediately before the survey took place, the southeast coast of the United States had experienced two very damaging hurri-

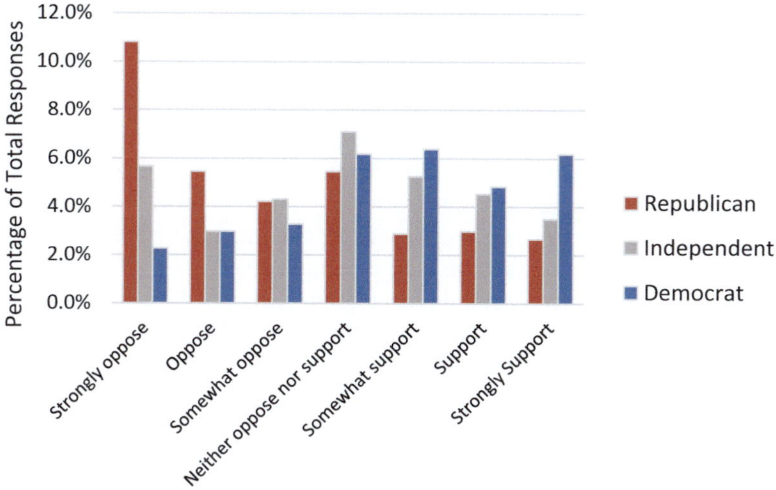

Fig. 8.5 Support for increasing the price of fossil fuels

canes with associated flooding, property loss, and death. The first was Hurricane Florence that made landfall near Wrightsville Beach, North Carolina on September 14 as a Category 1 hurricane, with 51 associated deaths and losses exceeding \$17 billion. Florida was even more directly affected by Hurricane Michael. Initially designated as Category 4 hurricane at the time of landfall on October 10, 2018, the National Hurricane Center later upgraded the storm to a Category 5 hurricane in a post-storm analysis that estimated sustained winds to have been 257.495 kph (160 mph). With the new classification, Hurricane Michael became only the fourth Category 5 hurricane to strike the U.S. coast in recorded history, and only the second known Category 5 hurricane to make landfall on the northern Gulf coast (National Oceanic and Atmospheric Administration 2019). Property damage was estimated at \$25 billion, and there were 46 deaths including 21 in Bay County, Florida.

It could be argued that the results from our experiment may have been "contaminated" by what experimental methods experts refer to as a *pretreatment effect* (Druckman and Leeper 2012; Gaines et al. 2007). This occurs when an event takes place immediately prior to an experiment that causes nearly all respondents to be exposed to the "experimental stimuli," such as information about damage that accompanies a major hurricane. In the case of our experiment, media coverage of Hurricane Michael immediately prior to when our survey was fielded may have resulted in respondents forming salient and strong opinions about the likely effects of sea-level rise, and the causes of increasingly intense hurricanes.

Any study outside of a controlled laboratory setting runs the risk of outside incidents influencing study participants. However, with respect to this study, there are two important points to be noted. First, although the hurricane made landfall in Florida, its path took it through the panhandle region of northwest Florida. The study area we surveyed was not directly affected by hurricane-force winds although there were several inches of rain in South Florida. Second, the information we presented to the survey respondents was very different from general news coverage of the two hurricanes.

Yet, even within this context, a sizeable proportion of the coastal South Florida residents, particularly those who identify themselves as Republican and conservative, are not yet convinced that sea-level rise threatens them. In addition, even Democrats and Independents are skeptical about the impacts that sea-level rise will have on their property values and on their communities.

The challenge of effectively communicating the reality of climate change remains immense despite the enormity of effort that has gone into studying various approaches and methods. The keys to effective communication of this now highly politicized topic must remain a high priority for social science research.

References

Allen G (2015) Florida Gov. Scott denies banning phrase 'climate change'. In: National Public Radio. Available at: https://www.npr.org/2015/03/10/392142452/florida-gov-scott-denies-banning-phrase-climate-change. Accessed 14 June 2018

Associated Press-NORC Center for Public Affairs Research at the University of Chicago (2018) Is the public willing to pay to help fix climate change? EPIC Energy Policy Institute at the University of Chicago. Available at: https://epic.uchicago.edu/sites/default/files/Full%20poll%20AP-NORC%202019.pdf. Accessed 18 Jan 2019

Balaji R, Kumar S, Misra A (2017) Understanding the effects of seawall construction using a combination of analytical modelling and remote sensing techniques: case study of Fansa, Gujarat, India. Int J Ocean Clim Sci Technol Impact 8(3):153–160. https://doi.org/10.1177/1759313117712180

Barszewski L (2018) Broward could require higher seawalls. In: South Florida Sun Sentinel. Available at: https://www.sun-sentinel.com/local/broward/fl-ne-army-corps-future-flooding-study-20180910-story.html. Accessed 20 Sept 2018

Building a Safer Florida Inc (2017) Revised coastal construction control line requirements. In: 6th edn Florida Building Code. Available at: http://www.floridabuilding.org/fbc/thecode/2017-6edition/BASF_2017_CCCL_061217.pdf

Center for Climate Strategies/Governor's Action Team on Energy & Climate Change (2008) Florida's Energy and Climate Change Action Plan: October 15, 2008. Available at: https://www.cakex.org/sites/default/files/documents/Florida's%20Action%20Plan%20on%20Energy%20and%20Climate%20Change_0.pdf

Congressional Budget Office (2018) Options for reducing the deficit: 2019–2028: Impose a tax on emissions of greenhouse gases. Available at: https://www.cbo.gov/budget-options/2018/54821. Accessed 16 Dec 2018

Cusick D (2018) FEMA approves buyout funds for Houston hoes flooded by Harvey. In: E&E News: Natural Disasters. Scientific American. Available at: https://www.scientificamerican.com/article/fema-approves-buyout-funds-for-houston-homes-flooded-by-harvey/. Accessed 11 June 2018

Druckman J, Leeper T (2012) Learning more from political communication experiments: pretreatment and its effects. Am J Polit Sci 56(4):875–896. https://doi.org/10.1111/j.1540-5907.2012.00582.x

EPA: United States Environmental Protection Agency (2017) Greenhouse Gas Emissions. Available at: https://www.epa.gov/ghgemissions. Accessed 28 Mar 2018

Flechas J (2017) Miami Beach to begin new $100 million flood prevention project in face of sea-level rise. In: Miami Herald. Available at: https://www.miamiherald.com/news/local/community/miami-dade/miami-beach/article129284119.html. Accessed 14 May 2018

Florida Department of Environmental Protection (2019) Office of Resilience and Coastal Protection. Available at: https://floridadep.gov/rcp. Accessed 22 May 2018

Gaines B, Kuklinski J, Quirk P (2007) The logic of the survey experiment reexamined. Polit Anal 15(1):1–20. https://doi.org/10.1093/pan/mpl008

Gleckman H (2018) Economists love carbon taxes. Voters Don't. In: Forbes. Available at: https://www.forbes.com/sites/howardgleckman/2018/12/27/economists-love-carbon-taxes-voters-dont/#a76241c4338d. Accessed 8 Jan 2019

Green A (2019) In Florida, a new governor speaks the words 'climate change'. In: WFSU Public Media News. Available at: https://news.wfsu.org/post/florida-new-governor-speaks-words-climate-change. Accessed 28 Apr 2019

Lynch M, Ozymy J, Jarrell M (2019) Executive actors and environmental enforcement: examining the "Rick Scott Effect" in the U.S. State of Florida. Rev Policy Res 36(3):395–413. https://doi.org/10.1111/ropr.12327

NOAA National Oceanic and Atmospheric Administration (2019) Hurricane Michael upgraded to a Category 5 at time of U.S. landfall: Post-storm analysis estimates sustained winds of 160 mph. Available at: https://www.noaa.gov/media-release/hurricane-michael-upgraded-to-category-5-at-time-of-us-landfall. Accessed 19 Apr 2019

Ruppert T, Deady E (2017) Climate change impacts on law and policy in Florida. In: Florida's climate: changes, variations, & impacts. https://doi.org/10.17125/fci2017.ch07

Southeast Florida Regional Compact (2017) Integrating the unified sea-level rise projection into local plans. In: Regional Climate Action Plan (RCAP) Workshop #10. 14 July 2016. Available at: http://southeastfloridaclimatecompact.org/wp-content/uploads/2017/01/SLRGuidance-Doc.pdf. Accessed 30 June 2018

Woolfenden T (2013) Sea walls designed to save beaches may actually speed up erosion. In: WLRN Miami-South Florida. Available at: https://www.wlrn.org/post/sea-walls-designed-save-beaches-may-actually-speed-erosion. Accessed 29 Aug 2018

Appendices

Appendix 1: Questionnaire

Section I Background

What is your age?
Please select your county of residence.
Are you male or female?

Male
Female

Which of the following do you consider to be your primary racial or ethnic group?

White
African American
Asian American
Hispanic
Other

What is your estimate of your family's annual household income (before taxes)?

Less than $30,000
$30,000–$69,999
$70,000–$99,999
$100,000–$200,000
Greater than $200,000

What is the highest level of education you have attained?

Less than a high school
High school graduate
Some college
4-year college degree
Advanced degree

R. Palm, T. Bolsen, *Climate Change and Sea Level Rise in South Florida*,
Coastal Research Library 34, https://doi.org/10.1007/978-3-030-32602-9

Which point on this scale best describes your political views?

Very conservative
Mostly conservative
Somewhat conservative
Moderate
Somewhat liberal
Mostly liberal
Very liberal

[If "Moderate" was the response to the previous question] Do you lean conservative, lean liberal, or consider yourself to be a pure moderate?

Lean Liberal
Lean Conservative
Pure Moderate

Generally speaking, which of the options on the scale below best describes your party identification?

Strong Republican
Weak Republican
Lean Republican
Independent
Lean Democrat
Weak Democrat
Strong Democrat

[If "Independent" was the response to the previous question] Do you lean Republican, lean Democrat, or consider yourself to be a pure Independent?

Lean Republican
Lean Democrat
Pure Independent

Some people think it is important to protect the environment even if it costs some jobs or otherwise reduces our standard of living. Other people think that protecting the environment is not as important as maintaining jobs and our standard of living. Which is closer to the way you feel?

Definitely protect environment
Very likely protect environment
Probably protect environment
Equally important
Probably maintain prosperous economy
Very likely maintain prosperous economy
Definitely maintain prosperous economy

To what extent do you disagree or agree that: "We have gone too far in pushing equal rights in this country"?

Strongly disagree
Moderately disagree
Slightly disagree
Neither agree nor disagree
Slightly agree
Moderately agree
Strongly agree

To what extent do you disagree or agree that: "If the government spent less time trying to fix everyone's problems, we'd all be a lot better off"?

Strongly disagree
Moderately disagree
Slightly disagree
Neither agree nor disagree
Slightly agree
Moderately agree
Strongly agree

To what extent do you think science enables us to overcome almost any problem?

Not at all
To a small extent
To a moderate extent
To a great extent
To a very great extent

In general, how interested are you in politics?

Extremely Disinterested
Disinterested
Somewhat Disinterested
Neither interested nor disinterested
Somewhat Interested
Interested
Extremely Interested

What type of housing do you live in?

Single family detached
Condo
Townhouse
Apartment
Multi-family

Do you own your home, or do you rent?

Rent
Own
(for owners only):

In what year did you purchase this house?

Did anyone, such as the real estate agent or the mortgage lender, discuss flooding with you during the process of purchasing your home?

Yes
No
Do not remember

Do you have a separate flood insurance policy in addition to your homeowner's policy?

Yes
No
Don't Know

How important were the following factors to you as you decided where you wanted to live?

Quality of local schools
Proximity to work
Proximity to restaurants, shops, and entertainment
Safety from flooding
Safety of the neighborhood from crime
Other

How long have you lived in Florida?

Many people don't know the answers to the next set of questions, so if you don't know the answer, just leave it blank or check "don't know."

Which gas makes up most of Earth's atmosphere?

Oxygen
Hydrogen
Carbon Dioxide
Nitrogen
Don't know

Seasons (summer, fall, winter, spring) are caused by the Earth being tilted on its axis.

True
False

Lasers work by focusing sound waves.

True
False

How long does it take for the Earth to go around the Sun?

1 day

1 month
1 year

Which kind of waves are used to make and receive cellphone calls?

Radio waves
Visible light waves
Sound waves
Gravity waves

What does a light-year measure?

Brightness
Time
Distance
Weight

Section II Treatment

Respondents in the treatment group were shown one of the maps below depending on their place of residence. All of the maps are from FloodIQ.com.

The map below shows the projected impact of flooding following a Category 3 hurricane on the Ft. Lauderdale area.

The map below shows the projected impact of flooding following a Category 3 hurricane on the Ft. Myers area.

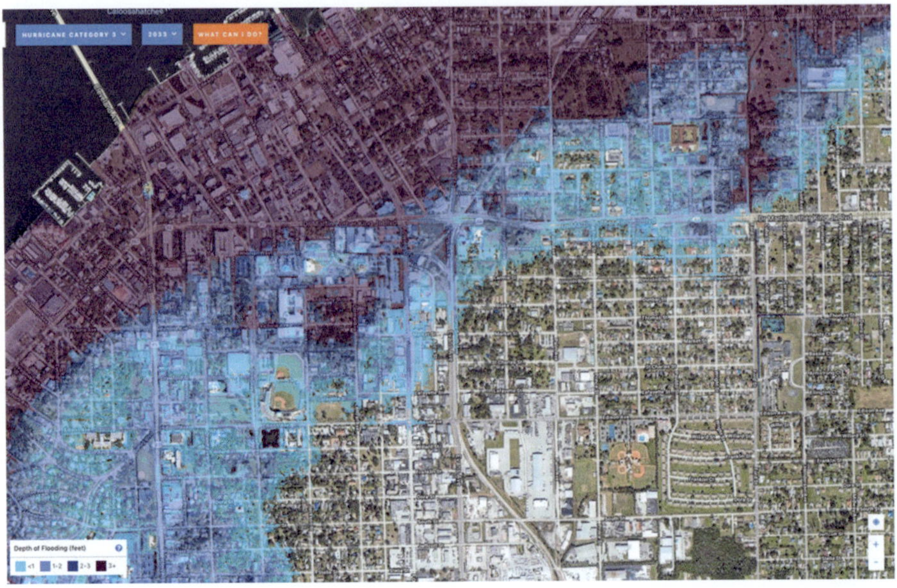

The map below shows the projected impact of flooding following a Category 3 hurricane on the Key West area.

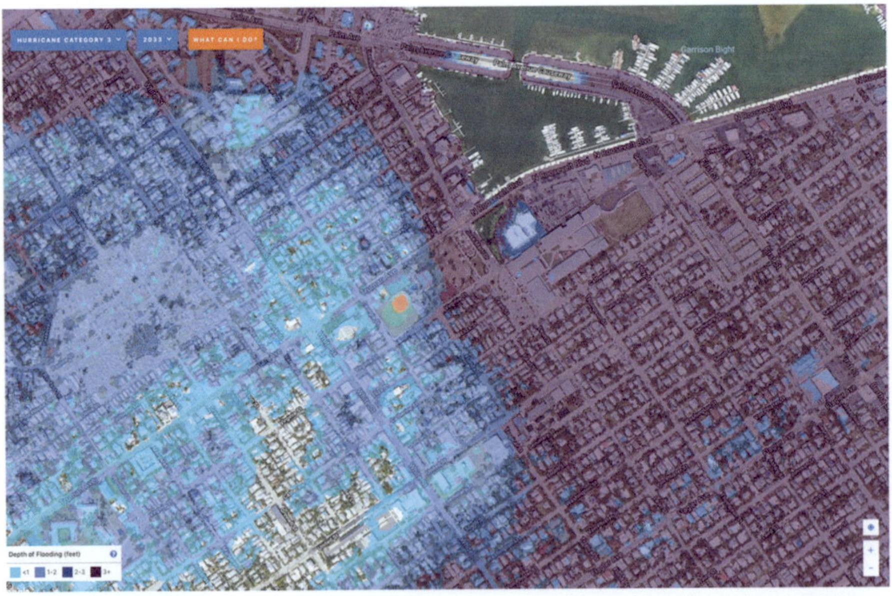

The map below shows the projected impact of flooding following a Category 3 hurricane on the Miami Dade area.

The map below shows the projected impact of flooding following a Category 3 hurricane on the St. Petersburg area.

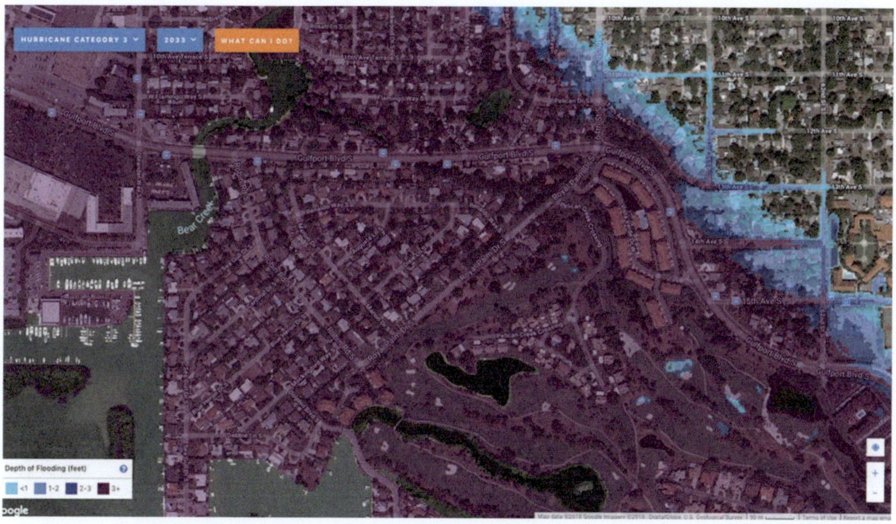

The map below shows the projected impact of flooding following a Category 3 hurricane on the Tampa area.

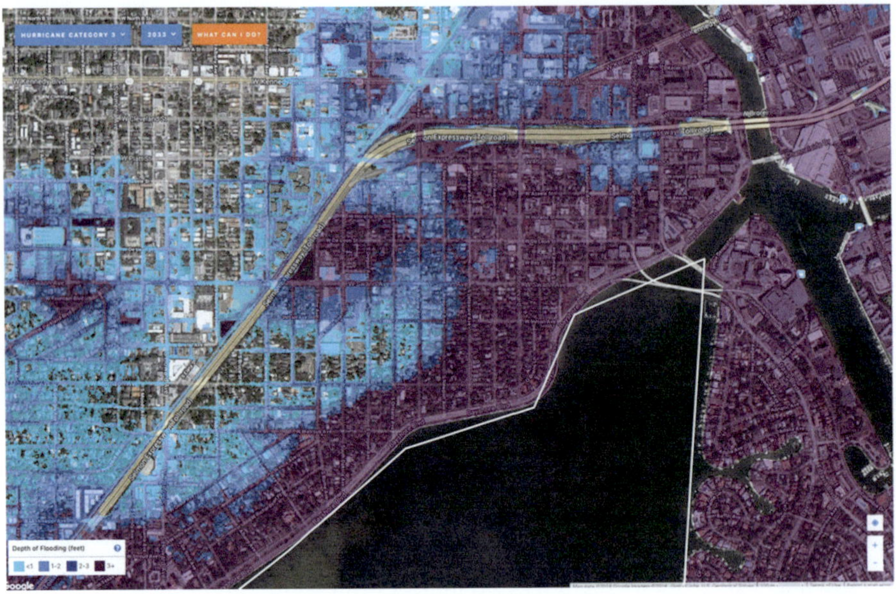

The map below shows the projected impact of flooding following a Category 3 hurricane on the Vero Beach area.

Section III Post-treatment Questions

To what extent do you believe that the sea-level is rising?

Definitely is NOT rising

Very likely is NOT rising
Probably is NOT rising
Not sure
Probably IS rising
Very likely IS rising
Definitely IS rising

Is your home susceptible to sea-level rise?

Definitely NOT
Very likely NOT
Probably NOT
Not sure
Probably YES
Very likely YES
Definitely YES

If sea-level rise occurs, to what extent will it reduce property values for the ____ area?

Not at all
A little
Somewhat
A moderate amount
A great deal

If sea-level rise occurs, to what extent will it reduce the property value of your own home?

Not at all
A little
Somewhat
A moderate amount
A great deal

Do you think that climate change is happening?

Definitely is NOT happening
Very likely is NOT happening
Probably is NOT happening
Not sure
Probably IS happening
Very likely IS happening
Definitely IS happening

To what extent do you think sea-level rise is related to climate change?

Definitely is NOT related to climate change
Very likely is NOT related to climate change
Probably is NOT related to climate change

Not sure
Probably IS related to climate change
Very likely IS related to climate change
Definitely IS related to climate change

To what extent do you believe that climate change is responsible for causing more extreme storms and hurricanes?

None at all
A little
A moderate amount
A lot
A great deal

How much do you support or oppose the following approaches to dealing with sea-level rise in Florida?
Change zoning laws to require greater set-backs from coast

Strongly oppose
Moderately oppose
Slightly oppose
Neither oppose nor support
Slightly support
Moderately support
Strongly support

How much do you support or oppose the following approaches to dealing with sea-level rise in Florida?
Government buys back land from people who live in areas that will flood.

Strongly oppose
Moderately oppose
Slightly oppose
Neither oppose nor support
Slightly support
Moderately support
Strongly support

How much do you support or oppose the following approaches to dealing with sea-level rise in Florida?
Government builds flood barriers.

Strongly oppose
Moderately oppose
Slightly oppose
Neither oppose nor support
Slightly support
Moderately support
Strongly support

Some companies burn fuels like oil, coal and natural gas, which produce greenhouse gases. Do you think the federal government should or should not tax these companies on the amount of greenhouse gases they put out?

Strongly oppose
Moderately oppose
Slightly oppose
Neither oppose nor support
Slightly support
Moderately support
Strongly support

To what extent do you support or oppose increasing the price of fossil fuels (like gasoline) to encourage people to save energy and encourage the development of energy efficient devices.

Strongly oppose
Moderately oppose
Slightly oppose
Neither oppose nor support
Slightly support
Moderately support
Strongly support

How concerned are you about the effects that climate change will have on you, your family, and your community?

Extremely unconcerned
Very unconcerned
Somewhat unconcerned
Neither concerned nor unconcerned
Somewhat concerned
Very concerned
Extremely concerned

As you think about sea-level rise, how much anxiety do you feel?

None at all
A little
A moderate amount
A lot
A great deal

Do you agree or disagree with the following statement: "I believe my actions have an influence on global warming and climate change."

Strongly disagree
Moderately disagree
Slightly disagree

Neither agree nor disagree
Slightly agree
Moderately agree
Strongly agree

Do you agree or disagree with the following statement: "It is governments, not citizens, that are primarily responsible for dealing with climate change."

Strongly disagree
Moderately disagree
Slightly disagree
Neither agree nor disagree
Slightly agree
Moderately agree
Strongly agree

How interested are you in learning more about the possible effects of sea-level rise?

Extremely uninterested
Very uninterested
Somewhat uninterested
Neither uninterested nor interested
Somewhat interested
Very interested
Extremely interested

Do you agree with the following statement: "Most scientific research on climate change is politically motivated."

Strongly disagree
Moderately disagree
Slightly disagree
Neither agree nor disagree
Slightly agree
Moderately agree
Strongly agree

Do you agree with the following statement: "Most scientists think climate change is happening primarily as a result of human activities"?

Strongly disagree
Moderately disagree
Slightly disagree
Neither agree nor disagree
Slightly agree
Moderately agree
Strongly agree

Do you agree with the following statement: "the idea that climate change is primarily due to human activities is a hoax or a conspiracy"?

Strongly disagree
Moderately disagree
Slightly disagree
Neither disagree nor agree
Slightly agree
Moderately agree
Strongly agree

Appendix 2: Technical Analysis

With respect to the belief that climate change is happening, we found that even with the inclusion of important independent variables that account for demographic and political characteristics, exposure to the localized flood map (i.e., "Treatment") was associated with significantly lowered the perception that climate change is happening ($p < .05$) (Table A1). Party identification, ideology, and other factors shape the opinions of South Florida residents. Republicans and those who identified as conservatives were less likely to believe that climate change is happening compared to Independents and Moderates ($p < .01$). Respondents who reported higher levels of education were more likely to believe climate change is happening ($p < .05$), as were those who possessed higher levels of factual scientific knowledge ($p < .01$). Females expressed a stronger belief in climate change compared to males ($p < .01$), and Liberals were more likely to say climate change is happening relative to Independents ($p < .05$). Individuals who prioritized the economic values over environmental values were also significantly less likely to report that climate change is happening ($p < .01$).

Model 2 in Table A1 reports the impact of demographic and political characteristics on respondents' beliefs that climate change is causing more extreme storms and hurricanes. First, similar to the unexpected results in Model 1, those exposured to the localized flood map were less likely to perceive that climate change is causing more extreme hurricanes and storms ($p < .01$). Higher levels of formal education are associated with the belief that climate change is causing more severe storms ($p < .05$); however, possessing higher levels of factual scientific information had no effect on beliefs about the climate change's relationship to the intensity of tropical storms. Females were more likely than males to view climate change as linked to an increase in the severity of hurricanes ($p < .01$), and respondents who reported higher levels of income were also less likely to link climate change to extreme weather ($p < .10$). Party identification, ideology, and personal values exerted the most powerful impact on respondents' beliefs about the link between climate change and the intensity of local storms. Republicans, conservatives, and those who prioritized the

Table A1 Beliefs about climate change and sea-level rise

	Happening	Storms	Sea-level climate	Sea-level happening	Sea-level house	Reduce value city	Reduce value house
	(Model 1)	(Model 2)	(Model 3)	(Model 4)	(Model 5)	(Model 6)	(Model 7)
Treatment	−0.18**	−0.15**	−0.07	−0.03	−0.23**	0.00	−0.21*
	(−0.08)	(−0.07)	(−0.08)	(−0.08)	(−0.11)	(−0.07)	(−0.12)
Age	0.00	0.00	0.01**	0.01***	−0.01***	0.00	0.00
	(0.00)	(0.00)	(0.00)	(0.00)	(0.00)	(0.00	(0.00)
Income	−0.05	−0.06*	−0.05	−0.06	0.16***	−0.03	−0.04
	(−0.04)	(−0.03)	(−0.04)	(−0.04)	(−0.06)	(−0.04)	(−0.06)
Education	0.10**	0.11***	0.16***	0.19***	0.12**	0.08**	0.09
	(−0.05)	(−0.04)	(−0.05)	(−0.05)	(−0.06)	(−0.04)	(−0.06)
Science knowledge	0.08***	0.01	0.10***	0.07***	0.08**	0.03	0.03
	(−0.03)	(−0.02)	(−0.03)	(−0.03)	(−0.04)	(−0.02)	(−0.04)
Female	0.35***	0.21***	0.22**	0.10	0.18	0.19**	−0.05
	(−0.09)	(−0.08)	(−0.10)	(−0.09)	(−0.13)	(−0.08)	(−0.14)
Minority	0.07	−0.11	−0.09	−0.05	−0.01	0.06	0.29*
	(−0.11)	(−0.10)	(−0.12)	(−0.11)	(−0.15)	(−0.10)	(−0.17)
Democrat	0.10	0.23**	0.10	0.08	−0.42***	0.06	−0.11
	(−0.11)	(−0.10)	(−0.12)	(−0.12)	(−0.16)	(−0.10)	(−0.18)
Republican	−0.52***	−0.43***	−0.44***	−0.36***	−0.46***	−0.17	−0.54***
	(−0.12)	(−0.10)	(−0.12)	(−0.12)	(−0.16)	(−0.10)	(−0.17)
Conservative	−0.52***	−0.46***	−0.58***	−0.43***	−0.16	−0.12	0.04
	(−0.11)	(−0.09)	(−0.11)	(−0.11)	(−0.15)	(−0.10)	(−0.16)
Liberal	0.27**	0.33***	0.38***	0.37***	0.32**	0.18*	−0.04
	(−0.12)	−0.10)	(−0.12)	(−0.12)	(−0.16)	(−0.10)	(−0.19)
Env. v Econ.	−0.30***	−0.27***	−0.29***	−0.27***	−0.21***	−0.14***	−0.16***
	(−0.03)	(−0.02)	(−0.03)	(−0.03)	(−0.03)	(−0.02)	(−0.04)
Constant	6.10***	4.35***	5.52***	5.28***	4.30***	3.62***	3.09***
	(−0.23)	(−0.20)	(−0.24)	(−0.24)	(−0.32)	(−0.21)	(−0.37)
R^2	0.34	0.38	0.33	0.29	0.11	0.09	0.09
N	971	971	971	971	971	971	541

Note: Cell entries are OLS coefficients with standard errors in parentheses below
*$p < 0.10$, **$p < 0.05$, ***$p < 0.01$

economy over the environment were significantly less likely to say that climate change is causing storms and hurricanes to become more extreme.

Model 3 and Model 4 in Table A1 report the factors associated with respondents' perceptions about the relationship between sea-level rise and climate change, and the degree to which sea-level rise is actually happening, respectively. Age, education, and science knowledge are positively associated with both beliefs. Females are more likely than males to believe sea-level rise is being caused by climate change. Partisanship, political ideology, and personal values again play the largest role in explaining respondents' beliefs about sea-level rise and its cause. For example, Model 3 in Table A1 reports a substantively large (i.e., approximately 1-point shift on the 7-point response scale) predicted difference between liberals' and conservatives' views on the linkage between climate change and sea-level rise. Similarly, individuals who prioritize economic growth over protecting the environment are more likely to reject the fact that the sea-level is rising and that this rise is linked to climate change.

Model 5 in Table A1 reports the factors that shaped respondents' perceptions that their own home is susceptible to flooding as a result of sea-level rise. As discussed earlier, those exposed to the maps were less likely to believe that their own home was susceptible to flooding that will result from sea-level rise, and this effect persists in the presence of the inclusion of additional variables that account for demographic and political characteristics of each respondent. Democrats and Republicans are *both* significantly less likely than Independents (the omitted baseline) to report that their own homes are susceptible to flooding that would result from sea-level rise; whereas liberals are significantly more likely than moderates to view their home as susceptible to flooding from sea-level rise. Prioritizing economic growth over protecting the environment is associated with lower perceptions that one's home may be flooded as a result of sea-level rise.

Model 6 and Model 7 in Table A1 report the factors that shaped respondents' perceptions about how sea-level rise would impact property values in their community and the value of their own homes. Perhaps the most striking result in both models is the almost nonexistent role that party identification and political ideology play in shaping views about how sea-level rise may affect property values. Party and ideology matter in only two instances across both Model 6 and Model 7 – that is, Liberals are more likely to believe sea-level rise presents a threat to property values of the community, and Republicans were more likely to believe that sea-level rise will not present a threat to the value of their own home. Respondents who value the environment over the economy are significantly more likely to believe that sea-level rise will reduce property values in their communities and the property value of their own home.

The survey included measures that we did not hypothesize that exposure to the localized flood map would influence, but nonetheless are important to consider in explaining beliefs about climate change and sea-level rise: (1) personal *concern* about the effects that sea-level rise will have on "you, your family, and your community" (1 = *extremely unconcerned*/7 = *extremely concerned*); (2) *anxiety* when thinking about sea-level rise 1 = *none at all*/7 = *a great deal*); (3) agreement with a statement that government (as opposed to individuals) is primarily responsible for dealing with climate change to tap *responsibility attributions* (1 = *strongly disagree*/7 = *strongly agree*); and, (4) agreement with a statement that "most scientists think climate change is happening primarily as a result of human activity" to assess perceptions that a scientific consensus exists (1 = *strongly disagree*/7 = *strongly agree*).

We report, in Table A2, the results from four ordinary least squares regressions used to evaluate the factors associated with greater concern and anxiety about sea-level rise, as well as beliefs about who is responsible for taking action to address the

Table A2 Concern, anxiety, attributions, and consensus beliefs

	Concern (Model 1)	Anxiety (Model 2)	Respons. Attribution (Model 3)	Scientific consensus (Model 4)
Treatment	−0.06	−0.06	0.06	0.12
	(0.09)	(0.07)	(0.12)	(0.10)
Age	−0.00	−0.01***	−0.01**	−0.00
	(0.00)	(0.00)	(0.00)	(0.00)
Income	−0.06	−0.05	−0.05	−0.11**
	(0.05)	(0.03)	(0.06)	(0.05)
Education	0.16***	0.14***	0.15**	0.15***
	(0.05)	(0.04)	(0.06)	(0.06)
Science knowledge	−0.03	−0.06***	−0.12***	0.05
	(0.03)	(0.02)	(0.04)	(0.03)
Female	0.37***	0.00	−0.24*	0.09
	(0.11)	(0.08)	(0.13)	(0.11)
Minority	0.04	0.09	−0.05	−0.32**
	(0.13)	(0.10)	(0.16)	(0.14)
Democrat	0.07	0.18*	0.38**	0.22
	(0.13)	(0.10)	(0.16)	(0.14)
Republican	−0.21	−0.11	−0.09	−0.20
	(0.13)	(0.10)	(0.16)	(0.14)
Conservative	−0.55***	−0.14	−0.05	−0.45***
	(0.13)	(0.09)	(0.16)	(0.14)
Liberal	0.21	0.28***	0.31*	0.20
	(0.13)	(0.10)	(0.16)	(0.14)
Env. v Econ.	−0.34***	−0.19***	−0.03	−0.26***
	(0.03)	(0.02)	(0.04)	(0.03)
Constant	5.56***	3.22***	3.89***	5.75***
	(0.27)	(0.20)	(0.33)	(0.29)
R^2	0.26	0.21	0.05	0.17
N	971	971	971	971

Note: Cell entries are OLS coefficients with standard errors in parentheses below
$^{*}p < 0.10$, $^{**}p < 0.05$, $^{***}p < 0.01$

problem, and perceptions regarding the existence of a scientific consensus on human-induced climate change. Model 1 in Table A2 reports that individuals with higher levels of education express more concern about the effects of sea-level rise on their communities and families. Women also expressed greater concern about the effects of sea-level rise compared to males. Respondents who identified as politically conservative expressed relatively lower levels of concern about the effects of sea-level rise. Those who prioritize the environment over the economy expressed higher levels of concern about the effects that will occur as a result of sea-level rise.

Model 2 in Table A2 reports the results from a regression of *Anxiety* related to sea-level rise on demographic and political characteristics. First, higher levels of education are again positively associated with anxiety about sea-level rise; however, higher levels of objective scientific knowledge are associated with lower levels of anxiety about sea-level rise. There is a negative association between *Age* and *Anxiety* about sea-level rise. Self-identified liberals and respondents who prioritized the environment over the economy also expressed more anxiety about the effects of sea-level rise.

Model 3 in Table A2 reports the correlates of beliefs that government, as opposed to individuals, is primarily responsible for dealing with climate change (positive values on the coefficients are associated with a belief that government is responsible while negative values indicate a belief that individuals are responsible). Identifying as a Democrat, or as a Liberal, increases the likelihood that a respondent believes government is primarily responsible for dealing with climate change. Higher levels of education are associated with the belief that government is primarily responsible for dealing with climate change. Age and science knowledge, conversely, are associated with the belief that individuals – rather than government – are primarily responsible for dealing with climate change. Model 4 in Table A2 reports a strong positive association between a respondent's level of education and belief that a scientific consensus exists, and a strong negative association between self-identifying as a conservative, or prioritizing the economy over the environment, in shaping perceptions regarding scientific consensus.

Table A3 replicates the seven OLS regression models that appear in Table 1 with the inclusion of *Anxiety* and *Scientific Consensus* as additional independent variables in each model. Across every model in Table A3, anxiety about the effects of sea-level rise and perceptions that a scientific consensus exists regarding human-induced climate change are positively associated with belief climate change is happening, perceptions that climate change is causing more severe storms and hurricanes, belief that climate change is causing sea-level rise, perception that sea-level rise is happening, belief that one's home is susceptible to sea-level rise, and beliefs that sea-level rise will reduce property values in the local community and one's own home value.

The survey also asked respondents their interest in learning more information (*Learn More*) about the possible effects of sea-level rise, measured on a seven-point scale (1 = extremely uninterested/7 = extremely interested). We regressed *Learn More* on the demographic and political characteristics included in the prior analyses, as well as two additional independent variables that accounted for attributions of responsibility for action (individuals versus government) and personal efficacy beliefs – that is the degree to which a person perceives their own actions as "have an influence" on collective outcomes (1 = *strongly disagree*/7 = *strongly agree*). The results reported in Table A4, Model 1, indicate that anxiety about sea-level rise, perceptions that a scientific consensus on climate change exists, age, belief that government is responsible for action, and belief that personal actions are efficacious are all factors that are positively associated with interest in learning more about the effects of sea-level rise. Conservatives and respondents who valued the economy over the environment were significantly less interested in learning more information about sea-level rise.

Table A3 Anxiety, consensus perceptions & climate beliefs

	Happening	Storms	Sea-level climate	Sea-level happening	Sea-level house	Reduce value city	Reduce value house
Age	0.01***	0.00	0.01***	0.01***	−0.01**	0.01**	0.01**
	(0.00)	(0.00)	(0.00)	(0.00)	(0.00)	(0.00)	(0.00)
Income	−0.02	−0.02	−0.01	−0.02	0.19***	−0.01	−0.03
	(0.04)	(0.03)	(0.04)	(0.04)	(0.05)	(0.03)	(0.06)
Education	0.03	0.04	0.08*	0.12***	0.06	0.03	0.03
	(0.04)	(0.03)	(0.04)	(0.04)	(0.06)	(0.04)	(0.06)
Sci. Know	0.07***	0.01	0.10***	0.07***	0.09**	0.04*	0.04
	(0.02)	(0.02)	(0.02)	(0.02)	(0.03)	(0.02)	(0.04)
Female	0.32***	0.18***	0.19**	0.08	0.16	0.17**	−0.04
	(0.09)	(0.07)	(0.09)	(0.09)	(0.12)	(0.08)	(0.13)
Minority	0.12	−0.06	−0.03	0.01	−0.00	0.08	0.29*
	(0.10)	(0.08)	(0.10)	(0.10)	(0.15)	(0.10)	(0.16)
Democrat	0.00	0.13	−0.00	−0.02	−0.5***	−0.02	−0.26
	(0.10)	(0.09)	(0.11)	(0.11)	(0.15)	(0.10)	(0.17)
Repub.	−0.44***	−0.35***	−0.36***	−0.28***	−0.38**	−0.11	−0.4***
	(0.11)	(0.09)	(0.11)	(0.11)	(0.15)	(0.10)	(0.16)
Conserva.	−0.38***	−0.32***	−0.42***	−0.28***	−0.06	−0.02	0.09
	(0.10)	(0.08)	(0.10)	(0.10)	(0.15)	(0.09)	(0.15)
Liberal	0.16	0.22**	0.26**	0.25**	0.21	0.08	−0.19
	(0.11)	(0.09)	(0.11)	(0.11)	(0.15)	(0.10)	(0.18)
Env. v Econ.	−0.19***	−0.15***	−0.17***	−0.16***	−0.1***	−0.05**	−.07**
	(0.02)	(0.02)	(0.02)	(0.02)	(0.04)	(0.02)	(0.03)
Anxiety	0.24***	0.29***	0.24***	0.25***	0.35***	0.28***	0.5***
	(0.04)	(0.03)	(0.04)	(0.04)	(0.05)	(0.03)	(0.05)
Sci. Con.	0.26***	0.24***	0.28***	0.26***	0.14***	0.14***	0.01
	(0.02)	(0.02)	(0.02)	(0.02)	(0.04)	(0.02)	(0.03)
Constant	3.76***	1.99***	3.08***	2.96***	2.24***	1.96***	1.52***
	(0.27)	(0.22)	(0.27)	(0.27)	(0.38)	(0.25)	(0.30)
R^2	0.45	0.52	0.45	0.41	0.18	0.21	0.21
N	971	971	971	971	971	971	541

Note: Cell entries are OLS coefficients with standard errors in parentheses below
$^{*}p < 0.10$, $^{**}p < 0.05$, $^{***}p < 0.01$

Table A4 Determinants of info-seeking and policy opinions

	Learn more (Model 1)	Policy support (Model 2)
Age	0.01**	0.00**
	(0.00)	(0.00)
Income	0.03	0.02
	(0.04)	(0.03)
Education	−0.06	0.10***
	(0.04)	(0.03)
Science knowledge	0.04	0.01
	(0.03)	(0.02)
Female	0.09	0.20***
	(0.09)	(0.07)
Minority	0.07	−0.14*
	(0.11)	(0.08)
Democrat	−0.10	0.09
	(0.11)	(0.08)
Republican	0.11	−0.12
	(0.11)	(0.08)
Conservative	−0.23**	−0.12
	(0.11)	(0.08)
Liberal	0.07	0.13
	(0.11)	(0.08)
Env. v Econ.	−0.09***	−0.08***
	(0.03)	(0.02)
Anxiety	0.34***	0.24***
	(0.04)	(0.03)
Scientific consensus	0.12***	0.14***
	(0.03)	(0.02)
Respons. Attribution	0.06***	0.07***
	(0.02)	(0.02)
Personal efficacy	0.29***	0.14***
	(0.03)	(0.02)
Constant	1.69***	1.93***
	(0.30)	(0.21)
R^2	0.44	0.49
N	971	971

Note: Cell entries are OLS coefficients with standard errors in parentheses below
$^{*}p < 0.10$, $^{**}p < 0.05$, $^{***}p < 0.0$

The survey also included a battery of questions designed to measure support for different policy approaches for dealing with sea-level rise in Florida, such as changes to zoning laws near the coast, government land buy backs, and constructing flood barriers; the survey also included two questions that measured (1) support for taxing companies that produce greenhouse gases and (2) increasing the price of fossil fuels to encourage people to save energy and promote energy efficiency. We created a single measure of *Policy Support* that combined each individual policy support item into a scaled index ($\alpha = .71$). Table A4 (Model 2) reports the results from a regression of *Policy Support* on the demographic, political, and theoretically important factors (i.e., efficacy perceptions, consensus perception, anxiety) known to shape individuals' support for action to address climate change. First, in line with the gateway belief model's predictions, perceptions that a scientific consensus exist on climate change is positively associated with support for policy action. Individuals who prioritize the environment over the economy are significantly more supportive of polices to address the impacts of sea-level rise and climate change. Respondents who believe that personal actions can make a difference in fighting climate change (Personal Efficacy) are more supportive of polices to combat sea-level rise and climate change. Individuals who believe government is responsible for addressing the problem are also, perhaps unsurprisingly, more supportive of policy action. Several demographic characteristics are also associated with greater support for policy action to address climate change in South Florida. Females are more supportive of steps to address the problem, and so are the more educated and older respondents in the sample, while non-whites are less supportive of policy action. Party identification and political ideology play a strong role in shaping support for *Policy Action* – that is, identical models that exclude *anxiety, scientific consensus, responsibility attributions, and personal efficacy* – indicate that the effect of partisanship and ideology is mediated through the impact that these factors have on more proximate forces that determine policy support.

MIX
Papier aus verantwortungsvollen Quellen
Paper from responsible sources
FSC® C105338

If you have any concerns about our products,
you can contact us on
ProductSafety@springernature.com

In case Publisher is established outside the EU,
the EU authorized representative is:
Springer Nature Customer Service Center GmbH
Europaplatz 3, 69115 Heidelberg, Germany

Printed by Libri Plureos GmbH
in Hamburg, Germany